FSA

Math Practice

Grade 4

Complete Content Review Plus
2 Full-length FSA Math Tests

Elise Baniam - Michael Smith

FSA Math Practice Grade 4

Published in the United State of America By

The Math Notion

Email: info@Mathnotion.com

Web: WWW.MathNotion.com

ISBN: 978-1-63620-012-5

About the Author

Elise Baniam has been a math instructor for over a decade now. She graduated in Mathematics. Since 2006, Elise has devoted his time to both teaching and developing exceptional math learning materials. As a Math instructor and test prep expert, Elise has worked with thousands of students. She has used the feedback of her students to develop a unique study program that can be used by students to drastically improve their math score fast and effectively.

- **SAT Math Workbook**
- **ACT Math Workbook**
- **ISEE Math Workbooks**
- **SSAT Math Workbooks**
- **many Math Education Workbooks**
- **and some Mathematics books ...**

As an experienced Math teacher, Mrs. Baniam employs a variety of formats to help students achieve their goals: she teaches students in large groups, and she provides training materials and textbooks through her website and through Amazon.

You can contact Elise via email at:

Elise@Mathnotion.com

Get the Targeted Practice You Need to Excel on the Math Section of the FSA Test Grade 4!

FSA Math Practice Book Grade 4 is **an excellent investment in your future** and the best solution for students who want to maximize their score and minimize study time. Practice is an essential part of preparing for a test and improving a test taker's chance of success. The best way to practice taking a test is by going through lots of FSA math questions.

High-quality mathematics instruction ensures that students become problem solvers. We believe all students can develop deep conceptual understanding and procedural fluency in mathematics. In doing so, through this math workbook we help our students grapple with real problems, think mathematically, and create solutions.

FSA Math Practice Book allows you to:

- Reinforce your strengths and improve your weaknesses
- Practice **2500+ realistic** FSA math practice questions
- Exercise math problems in a variety of formats that provide intensive practice
- Review and study **Two Full-length FSA Practice Tests** with detailed explanations

...and much more!

This Comprehensive FSA Math Practice Book is carefully designed to provide only that **clear and concise information** you need.

WWW.MathNotion.com

… So Much More Online!

✓ FREE Math Lessons

✓ More Math Learning Books!

✓ Mathematics Worksheets

✓ Online Math Tutors

For a PDF Version of This Book

Please Visit WWW.MathNotion.com

Contents

Chapter 1:
Place Value and
Number Sense

Numbers in Standard Form

Write the number in standard form.

1) 10 million 208 thousand 24

_____

2) 72 million 9 thousand 708

_____

3) 121 million 24 thousand 453

_____

4) 541 million 75 thousand 127

_____

5) 90 billion 15 million 68 thousand 15

_____

6) 12 billion 120 million 5

_____

7) 8 billion 114 million 88 thousand

_____

8) 16 billion 28 thousand 785

_____

9) 75 billion 159 thousand 324

_____

10) 41 billion 3 million 8 thousand 25

_____

11) 16 billion 129 thousand 989

_____

12) 65 billion 220 million 6 thousand 2

_____

13) 785 million 124 thousand 97

_____

14) 33 billion 104 million 11 thousand 57

_____

15) 95 billion 424 million

_____

16) 27 billion 77 million 9 thousand 150

_____

Number in Expand Form

Write the number in expand form.

1) 956: _____ .

2) 3,800: _____ .

3) 52,457: _____ .

4) 60,070: _____ .

5) 409,389: _____ .

6) 76,805: _____ .

7) 745,321: _____ .

8) 8,146: _____ .

9) 19,037: _____ .

10) 52,799: _____ .

11) 5,125: _____ .

12) 400,544: _____ .

13) 600,700: _____ .

14) 3,080,000: _____ .

Odd or Even

Write odd or even.

1) 19 _____

2) 81 _____

3) 456 _____

4) 852 _____

5) 953 _____

6) 183 _____

7) 987 _____

8) 540 _____

9) 777 _____

10) 544 _____

11) 33 _____

12) 4,458 _____

13) 15,159 _____

14) 9,357_____

15) 3,000 _____

16) 14 _____

17) 257 _____

18) 660 _____

19) 45,789 _____

20) 15,300 _____

21) 452 _____

22) 49,459 _____

23) 84 _____

24) 7,700 _____

25) 6,451 _____

26) 985 _____

Compare Whole Numbers

Compare, writing <; >, or = between the numbers.

1) 40,420 ☐ 41,004

2) 29,460 ☐ 29,640

3) 78,920 ☐ 87,290

4) 34,570 ☐ 33,750

5) 96,328 ☐ 96,238

6) 85,843 ☐ 85,840

7) 76,584 ☐ 76,854

8) 72,998 ☐ 72,989

9) 37,467 ☐ 37,567

10) 48,878 ☐ 49,878

11) 56,660 ☐ 65,660

12) 73,898 ☐ 69,899

13) 89,990 ☐ 98,110

14) 84,760 ☐ 84,670

15) 26,680 ☐ 26,860

16) 86,440 ☐ 86,440

17) 158,980 ☐ 158,890

18) 201,807 ☐ 201,807

19) 243,240 ☐ 243,420

20) 345,566 ☐ 354,655

21) 187,158 ☐ 196,001

22) 137,983 ☐ 137,895

23) 278,788 ☐ 249,988

24) 194,854 ☐ 194,845

25) 219,390 ☐ 291,110

26) 305,288 ☐ 299,999

27) 317,857 ☐ 371,857

28) 405,710 ☐ 405,170

Pattern

Continue this pattern for four more numbers:

1) 1,400; 1,250; 1,100; 950; _____

2) 3,700; 3,500; 3,300; 3,100; _____

3) 4,200; 3,850; 3,500; 3,150; _____

4) 1,650; 1,530; 1,410; 1,290; _____

5) 2,900; 2,650; 2,400; 2,150; _____

6) 5,000; 4,600; 4,200; 3,800; _____

7) 3,950; 3,700; 3,450; 3,200; _____

8) 2,800; 2,675; 2,550; 2,425; _____

9) 3,850; 3,550; 3,250; 2,950; _____

10) 4,700; 4,500; 4,300; 4,100; _____

11) Write a list of five numbers that follows this pattern: Start at 250 and add 200

each time.

_____

Round Whole Numbers

Round to the place of the underlined digit.

1) 7,46<u>7</u>,589 ≈ _____

2) 54<u>6</u>,125 ≈ _____

3) 9,18<u>7</u>,208 ≈ _____

4) 15,68<u>5</u>,807 ≈ _____

5) 5,45<u>4</u>,676 ≈ _____

6) 3,58<u>8</u>,975 ≈ _____

7) 8,368,<u>5</u>19 ≈ _____

8) 27,754,<u>7</u>69≈ _____

9) 42,654,<u>4</u>11≈ _____

10) 7, <u>6</u>21,879 ≈ _____

11) 19,78<u>8</u>,987 ≈ _____

12) 4,28<u>6</u>,850 ≈ _____

13) 9,2<u>7</u>3,778 ≈ _____

14) 6,48<u>4</u>,684 ≈ _____

15) 5,157,<u>6</u>28 ≈ _____

16) 8,66<u>7</u>,885 ≈ _____

17) 3,56<u>7</u>,980 ≈ _____

18) 8,369,4<u>3</u>2 ≈ _____

19) 24,25<u>6</u>,880 ≈ _____

20) 5,2<u>2</u>9,758 ≈ _____

21) 6,9<u>8</u>7,422 ≈ _____

22) 4,87<u>7</u>,391 ≈ _____

Answer key Chapter 1

Numbers in Standard Form

1) 10,208,024

2) 72,009,708

3) 121,024,453

4) 541,075,127

5) 90,015,068,015

6) 12,120,000,005

7) 8,114,088,000

8) 16,000,028,785

9) 75,000,159,324

10) 41,003,008,025

11) 16,000,129,989

12) 65,220,006,002

13) 785,124,097

14) 33,104,011,057

15) 95,424,000,000

16) 27,077,009,150

Numbers in Expand Form

1) $(9 \times 100) + (5 \times 10) + 6$

2) $(3 \times 1,000) + (8 \times 100)$

3) $(5 \times 10,000) + (2 \times 1,000) + (4 \times 100) + (5 \times 10) + 7$

4) $(6 \times 10,000) + (7 \times 10) + 0$

5) $(4 \times 100,000) + (9 \times 1,000) + (3 \times 100) + (8 \times 10) + 9$

6) $(7 \times 10,000) + (6 \times 1,000) + (8 \times 100) + 5$

7) $(7 \times 100,000) + (4 \times 10,000) + (5 \times 1,000) + (3 \times 100) + (2 \times 10) + 1$

8) $(8 \times 1,000) + (1 \times 100) + (4 \times 10) + 6$

9) $(1 \times 10,000) + (9 \times 1,000) + (3 \times 10) + 7$

10) $(5 \times 10,000) + (2 \times 1,000) + (7 \times 100) + (9 \times 10) + 9$

11) $(5 \times 1,000) + (1 \times 100) + (2 \times 10) + 5$

12) $(4 \times 100,000) + (5 \times 100) + (4 \times 10) + 4$

13) $(6 \times 100,000) + (7 \times 100)$

14) $(3 \times 1,000,000) + (8 \times 10,000)$

Odd or Even

1) Odd

2) Odd

3) Even

4) Even

5) Odd

6) Odd

7) Odd

8) Even

9) Odd

10) Even

11) Odd

12) Even

13) Odd

14) Odd

15) Even

16) Even

17) Odd

18) Even

19) Odd

20) Even

21) Even

22) Odd

23) Even

24) Even

25) Odd

26) Odd

Compare Whole Numbers

1) <

2) <

3) <

4) >

5) >

6) >

7) <

8) >

9) <

10) <

11) <

12) >

13) <

14) >

15) <

16) =

17) >

18) =

19) <

20) <

21) <

22) >

23) >

24) >

25) <

26) >

27) <

28) >

Pattern

1) 800; 650; 500; 350

2) 2,900; 2,700; 2,500; 2,300

3) 2,800; 2,450; 2,100; 1,750

4) 1,170; 1,050; 930; 810

5) 1,900; 1,650; 1,400; 1,150

6) 3,400; 3,000; 2,600; 2,200

7) 2,950; 2,700; 2,450; 2,200

8) 2,300; 2,175; 2,050; 1,925

9) 2,650; 2,350; 2,050; 1,750

10) 3,900; 3,700; 3,500; 3,300

11) 250; 450; 650; 850; 1,050

Round whole number

1) 7,470,000

2) 546,000

3) 9,187,000

4) 15,686,000

5) 5,455,000

6) 3,589,000

7) 8,368,500

8) 27,754,800

9) 42,654,400

10) 7,600,000

11) 19,789,000

12) 4,287,000

13) 9,270,000

14) 6,485,000

15) 5,157,600

16) 8,668,000

17) 3,568,000

18) 8,369,430

19) 24,257,000

20) 5,230,000

21) 6,990,000

22) 4,877,000

Chapter 2: Adding and Subtracting

Adding 3-Digit Numbers

Find each sum.

$$
\begin{array}{r}
526 \\
1) \quad + \ 236 \\
\hline
\end{array}
$$

$$
\begin{array}{r}
689 \\
7) \quad + \ 456 \\
\hline
\end{array}
$$

$$
\begin{array}{r}
469 \\
13) + 156 \\
\hline
\end{array}
$$

$$
\begin{array}{r}
725 \\
2) \quad + \ 130 \\
\hline
\end{array}
$$

$$
\begin{array}{r}
863 \\
8) \quad + \ 325 \\
\hline
\end{array}
$$

$$
\begin{array}{r}
360 \\
14) + 150 \\
\hline
\end{array}
$$

$$
\begin{array}{r}
425 \\
3) \quad + \ 153 \\
\hline
\end{array}
$$

$$
\begin{array}{r}
965 \\
9) \quad + \ 865 \\
\hline
\end{array}
$$

$$
\begin{array}{r}
689 \\
15) + 263 \\
\hline
\end{array}
$$

$$
\begin{array}{r}
563 \\
4) \quad + \ 125 \\
\hline
\end{array}
$$

$$
\begin{array}{r}
369 \\
10) + \ 120 \\
\hline
\end{array}
$$

$$
\begin{array}{r}
890 \\
16) + \ 345 \\
\hline
\end{array}
$$

$$
\begin{array}{r}
453 \\
5) \quad + \ 230 \\
\hline
\end{array}
$$

$$
\begin{array}{r}
187 \\
11) + 125 \\
\hline
\end{array}
$$

$$
\begin{array}{r}
720 \\
17) + 215 \\
\hline
\end{array}
$$

$$
\begin{array}{r}
398 \\
6) \quad + \ 120 \\
\hline
\end{array}
$$

$$
\begin{array}{r}
389 \\
12) + 150 \\
\hline
\end{array}
$$

$$
\begin{array}{r}
680 \\
18) + 230 \\
\hline
\end{array}
$$

Adding 4–Digit Numbers

Add.

1) $\begin{array}{r} 2,135 \\ + 5,236 \\ \hline \end{array}$

4) $\begin{array}{r} 3,125 \\ +4,035 \\ \hline \end{array}$

7) $\begin{array}{r} 3,236 \\ +2,369 \\ \hline \end{array}$

2) $\begin{array}{r} 4,369 \\ + 1,356 \\ \hline \end{array}$

5) $\begin{array}{r} 4,135 \\ +2,194 \\ \hline \end{array}$

8) $\begin{array}{r} 6,320 \\ +3,765 \\ \hline \end{array}$

3) $\begin{array}{r} 6,598 \\ + 2,325 \\ \hline \end{array}$

6) $\begin{array}{r} 5,036 \\ +2,365 \\ \hline \end{array}$

9) $\begin{array}{r} 3,890 \\ +3,567 \\ \hline \end{array}$

Find the missing numbers.

10) $1,155 + \underline{\quad} = 1,469$

13) $555 + \underline{\quad} = 1,886$

11) $400 + 3,000 = \underline{\quad}$

14) $\underline{\quad} + 920 = 1,550$

12) $5,200 + \underline{\quad} = 7,300$

15) $\underline{\quad} + 2,670 = 4,230$

16) $689,505 = 80,000 + 600,000 + 5 + \underline{\hspace{2cm}} + 9,000$

17) $750,678 = 50,000 + 700,000 + 8 + \underline{\hspace{2cm}} + 600$

18) $574,962 = 70,000 + 500,000 + 2 + \underline{\hspace{2cm}} + 900 + 60$

Estimate Sums

Estimate the sum by rounding each added to the nearest ten.

1) $36 + 9 =$

2) $29 + 46 =$

3) $36 + 12 =$

4) $37 + 38 =$

5) $12 + 35 =$

6) $38 + 13 =$

7) $48 + 25 =$

8) $36 + 77 =$

9) $45 + 86 =$

10) $62 + 58 =$

11) $45 + 36 =$

12) $52 + 18 =$

13) $35 + 59 =$

14) $38 + 65 =$

15) $87 + 82 =$

16) $18 + 69 =$

17) $65 + 64 =$

18) $33 + 26 =$

19) $73 + 48 =$

20) $35 + 64 =$

21) $13 + 93 =$

22) $63 + 52 =$

23) $164 + 142 =$

24) $54 + 77 =$

Subtracting 3–Digit Numbers

Find the difference.

1) $\begin{array}{r} 756 \\ -\ 236 \\ \hline \end{array}$

2) $\begin{array}{r} 693 \\ -\ 130 \\ \hline \end{array}$

3) $\begin{array}{r} 425 \\ -\ 153 \\ \hline \end{array}$

4) $\begin{array}{r} 365 \\ -\ 125 \\ \hline \end{array}$

5) $\begin{array}{r} 493 \\ -\ 230 \\ \hline \end{array}$

6) $\begin{array}{r} 398 \\ -\ 120 \\ \hline \end{array}$

7) $\begin{array}{r} 989 \\ -\ 756 \\ \hline \end{array}$

8) $\begin{array}{r} 863 \\ -\ 325 \\ \hline \end{array}$

9) $\begin{array}{r} 965 \\ -\ 465 \\ \hline \end{array}$

10) $\begin{array}{r} 369 \\ -\ 120 \\ \hline \end{array}$

11) $\begin{array}{r} 159 \\ -\ 125 \\ \hline \end{array}$

12) $\begin{array}{r} 789 \\ -\ 450 \\ \hline \end{array}$

13) $\begin{array}{r} 469 \\ -\ 156 \\ \hline \end{array}$

14) $\begin{array}{r} 960 \\ -\ 250 \\ \hline \end{array}$

15) $\begin{array}{r} 689 \\ -\ 358 \\ \hline \end{array}$

16) $\begin{array}{r} 890 \\ -\ 345 \\ \hline \end{array}$

17) $\begin{array}{r} 929 \\ -\ 115 \\ \hline \end{array}$

18) $\begin{array}{r} 999 \\ -\ 130 \\ \hline \end{array}$

Subtracting 4–Digit Numbers

Subtract.

1) $\begin{array}{r} 3,130 \\ -\ 1,134 \\ \hline \end{array}$

4) $\begin{array}{r} 6,987 \\ -\ 6,422 \\ \hline \end{array}$

7) $\begin{array}{r} 8,356 \\ -\ 5,712 \\ \hline \end{array}$

2) $\begin{array}{r} 3,356 \\ -\ 2,870 \\ \hline \end{array}$

5) $\begin{array}{r} 5,362 \\ -\ 3,331 \\ \hline \end{array}$

8) $\begin{array}{r} 8,350 \\ -\ 2,729 \\ \hline \end{array}$

3) $\begin{array}{r} 5,986 \\ -\ 2,678 \\ \hline \end{array}$

6) $\begin{array}{r} 7,365 \\ -\ 2,212 \\ \hline \end{array}$

9) $\begin{array}{r} 6,117 \\ -1,216 \\ \hline \end{array}$

Find the missing number.

10) $4,223 - \underline{\quad} = 2,320$

11) $5,856 - \underline{\quad} = 4,245$

12) $1,136 - 689 = \underline{\quad}$

13) $4,200 - \underline{\quad} = 2,450$

14) $5,870 - 2,650 = \underline{\quad}$

15) $6,360 - 4,320 = \underline{\quad}$

16) $8,165 - \underline{\quad\quad} = 4,303$

17) $5,060 - 1,867 = \underline{\quad}$

18) Bob had \$3,486 invested in the stock market until he lost \$2,198 on those investments. How much money does he have in the stock market now?

Estimate Differences

Estimate the difference by rounding each number to the nearest ten.

1) $58 - 23 =$

2) $34 - 24 =$

3) $75 - 48 =$

4) $43 - 24 =$

5) $69 - 46 =$

6) $42 - 23 =$

7) $77 - 47 =$

8) $49 - 28 =$

9) $94 - 48 =$

10) $79 - 59 =$

11) $68 - 26 =$

12) $83 - 37 =$

13) $73 - 43 =$

14) $58 - 42 =$

15) $82 - 52 =$

16) $65 - 43 =$

17) $99 - 81 =$

18) $42 - 24 =$

19) $58 - 47 =$

20) $89 - 28 =$

21) $81 - 65 =$

22) $68 - 14 =$

23) $76 - 6 =$

24) $78 - 31 =$

Subtract from Whole Thousands.

Find the difference.

1) $3,000 - 10 =$ ____

2) $4,000 - 5 =$ ____

3) $2,000 - 8 =$ ____

4) $5,000 - 30 =$ ____

5) $7,000 - 7 =$ ____

6) $6,000 - 15 =$ ____

7) $8,000 - 40 =$ ____

8) $9,000 - 5 =$ ____

9) $2,000 - 8 =$ ____

10) $5,000 - 30 =$ ____

11) $7,000 - 200 =$ ____

12) $6,000 - 2 =$ ____

13) $4,000 - 20 =$ ____

14) $8,000 - 200 =$ ____

15) $5,000 - 100 =$ ____

16) $6,000 - 80 =$ ____

17) $5,000 - 70 =$ ____

18) $7,000 - 200 =$ ____

19) $9,000 - 300 =$ ____

20) $2,000 - 8 =$ ____

21) $4,000 - 10 =$ ____

22) $8,000 - 50 =$ ____

23) $3,000 - 90 =$ ____

24) $1,000 - 6 =$ ____

25) $5,000 - 5 =$ ____

26) $8,000 - 90 =$ ____

27) $9,000 - 30 =$ ____

28) $2,000 - 60 =$ ____

Answer key Chapter 2

Adding three–digit numbers

1) 762	7) 1,145	13) 625
2) 855	8) 1,188	14) 510
3) 578	9) 1,830	15) 952
4) 688	10) 489	16) 1,235
5) 683	11) 312	17) 935
6) 518	12) 539	18) 910

Adding 4–digit numbers

1) 7,371	7) 5,605	13) 1,331
2) 5,725	8) 10,085	14) 630
3) 8,923	9) 7,457	15) 1,560
4) 7,160	10) 314	16) 500
5) 6,329	11) 3,400	17) 70
6) 7,401	12) 2,100	18) 4,000

Estimate sums

1) 50	7) 80	13) 100	19) 120
2) 80	8) 120	14) 110	20) 100
3) 50	9) 140	15) 170	21) 100
4) 80	10) 120	16) 90	22) 110
5) 50	11) 90	17) 130	23) 300
6) 50	12) 70	18) 60	24) 130

Subtracting 3–digit numbers

1) 520	7) 233	13) 313
2) 563	8) 538	14) 710
3) 272	9) 500	15) 331
4) 240	10) 249	16) 545
5) 263	11) 34	17) 814
6) 278	12) 339	18) 869

Subtracting 4–digit numbers

1) 1,996	2) 486	3) 3,308

4) 565	9) 4,901	14) 3,220
5) 2,031	10) 1,903	15) 2,040
6) 5,153	11) 1,611	16) 3,862
7) 2,644	12) 447	17) 3,193
8) 5,621	13) 1,750	18) 1,288

Estimate differences

1) 40	7) 30	13) 30	19) 10
2) 10	8) 20	14) 20	20) 60
3) 30	9) 40	15) 30	21) 10
4) 20	10) 20	16) 30	22) 60
5) 20	11) 40	17) 20	23) 70
6) 20	12) 40	18) 20	24) 50

Subtract from Whole Thousands

1) 2,990	11) 6,800	21) 3,990
2) 3,995	12) 5,998	22) 7,950
3) 1,992	13) 3,980	23) 2,910
4) 4,970	14) 7,800	24) 994
5) 6,993	15) 4,900	25) 4,995
6) 5,985	16) 5,920	26) 7,910
7) 7,960	17) 5,930	27) 8,970
8) 8,995	18) 6,800	28) 1,940
9) 1,992	19) 8,700	
10) 4,970	20) 1,992	

Chapter 3: Multiplication and Division

Multiplication Whole Numbers

Find the answers.

1) $\begin{array}{r} 53 \\ \times\ 12 \\ \hline \\ \hline \end{array}$

6) $\begin{array}{r} 45 \\ \times\ 21 \\ \hline \\ \hline \end{array}$

11) $\begin{array}{r} 363 \\ \times\ 4 \\ \hline \\ \hline \end{array}$

2) $\begin{array}{r} 46 \\ \times\ 10 \\ \hline \\ \hline \end{array}$

7) $\begin{array}{r} 12 \\ \times\ 13 \\ \hline \\ \hline \end{array}$

12) $\begin{array}{r} 36 \\ \times\ 20 \\ \hline \\ \hline \end{array}$

3) $\begin{array}{r} 17 \\ \times\ 12 \\ \hline \\ \hline \end{array}$

8) $\begin{array}{r} 42 \\ \times\ 20 \\ \hline \\ \hline \end{array}$

13) $\begin{array}{r} 345 \\ \times\ 23 \\ \hline \\ \hline \end{array}$

4) $\begin{array}{r} 45 \\ \times\ 14 \\ \hline \\ \hline \end{array}$

9) $\begin{array}{r} 140 \\ \times\ 7 \\ \hline \\ \hline \end{array}$

14) $\begin{array}{r} 725 \\ \times\ 30 \\ \hline \\ \hline \end{array}$

5) $\begin{array}{r} 48 \\ \times\ 12 \\ \hline \\ \hline \end{array}$

10) $\begin{array}{r} 564 \\ \times\ 4 \\ \hline \\ \hline \end{array}$

15) $\begin{array}{r} 364 \\ \times\ 25 \\ \hline \\ \hline \end{array}$

16) Emily has 17 candy bars. She divided each bar into 7 equal pieces to share with her colleagues. How many colleagues does Emily have? _____

17) Harper packaged cupcake in boxes of 12. She filled 36 boxes. How many cupcakes does Harper have? _____

Multiply Tens and Hundreds.

Multiply, and find the missing factors.

1) $50 \times 6 =$ _____

2) $7 \times 400 =$ _____

3) $30 \times 9 =$ _____

4) $80 \times 70 =$ _____

5) $6 \times 400 =$ _____

6) $70 \times 90 =$ _____

7) $20 \times 600 =$ _____

8) $10 \times 900 =$ _____

9) $60 \times 800 =$ _____

10) $7 \times 700 =$ _____

11) _____ $\times 4 = 320$

12) _____ $\times 8 = 6,400$

13) _____ $\times 7 = 2,100$

14) _____ $\times 9 = 720$

15) _____ $\times 3 = 3,600$

16) _____ $\times 600 = 5,400$

17) _____ $\times 70 = 2,800$

18) _____ $\times 30 = 2,700$

19) _____ $\times 200 = 1,400$

20) _____ $\times 300 = 12,000$

21) $90 \times$ _____ $= 4,500$

22) $40 \times$ _____ $= 2,400$

23) $80 \times$ _____ $= 8,000$

24) $60 \times$ _____ $= 420$

25) $30 \times$ _____ $= 15,000$

26) $700 \times$ _____ $= 6,3000$

27) $50 \times$ _____ $= 3,000$

28) $300 \times$ _____ $= 18,000$

Estimate Products

Estimate the products.

1) $38 \times 17 =$

2) $13 \times 16 =$

3) $23 \times 16 =$

4) $23 \times 12 =$

5) $65 \times 21 =$

6) $38 \times 71 =$

7) $42 \times 92 =$

8) $15 \times 39 =$

9) $23 \times 14 =$

10) $73 \times 33 =$

11) $43 \times 24 =$

12) $49 \times 13 =$

13) $58 \times 33 =$

14) $82 \times 56 =$

15) $52 \times 77 =$

16) $26 \times 58 =$

17) $34 \times 38 =$

18) $33 \times 47 =$

19) $32 \times 36 =$

20) $35 \times 47 =$

21) $75 \times 53 =$

22) $29 \times 11 =$

23) $53 \times 11 =$

24) $94 \times 36 =$

Multiplication Missing Numbers

Find the missing numbers.

1) $20 \times \underline{\quad} = 80$

2) $16 \times \underline{\quad} = 48$

3) $\underline{\quad} \times 12 = 96$

4) $12 \times \underline{\quad} = 48$

5) $\underline{\quad} \times 17 = 102$

6) $15 \times \underline{\quad} = 135$

7) $6 \times \underline{\quad} = 48$

8) $80 \times \underline{\quad} = 2,400$

9) $12 \times 7 = \underline{\quad}$

10) $36 \times 5 = \underline{\quad}$

11) $22 \times 4 = \underline{\quad}$

12) $69 \times 3 = \underline{\quad}$

13) $\underline{\quad} \times 45 = 270$

14) $9 \times \underline{\quad} = 360$

15) $70 \times \underline{\quad} = 280$

16) $32 \times \underline{\quad} = 256$

17) $\underline{\quad} \times 30 = 270$

18) $25 \times 5 = \underline{\quad}$

19) $\underline{\quad} \times 13 = 169$

20) $19 \times \underline{\quad} = 228$

21) $40 \times 6 = \underline{\quad}$

22) $50 \times 3 = \underline{\quad}$

23) $\underline{\quad} \times 26 = 832$

24) $18 \times \underline{\quad} = 216$

25) Emily has 24 candy bars. She divided each bar into 7 equal pieces to share with her colleagues. How many colleagues does Emily have? _____

Long Division by One Digit

Find the quotient.

1) $5\overline{)100}=$

2) $8\overline{)64}=$

3) $13\overline{)169}=$

4) $3\overline{)24}=$

5) $12\overline{)144}=$

6) $8\overline{)48}=$

7) $2\overline{)12}=$

8) $7\overline{)21}=$

9) $9\overline{)468}=$

10) $5\overline{)30}=$

11) $4\overline{)36}=$

12) $13\overline{)65}=$

13) $8\overline{)56}=$

14) $9\overline{)90}=$

15) $8\overline{)112}=$

16) $24\overline{)360}=$

17) $2\overline{)36}=$

18) $8\overline{)24}=$

19) $4\overline{)60}=$

20) $9\overline{)153}=$

21) $6\overline{)114}=$

22) $5\overline{)90}=$

23) $10\overline{)1,170}=$

24) $11\overline{)462}=$

25) $4\overline{)540}=$

26) $8\overline{)640}=$

27) $8\overline{)216}=$

28) $8\overline{)112}=$

29) $15\overline{)495}=$

30) $20\overline{)400}=$

31) $11\overline{)484}=$

32) $10\overline{)800}=$

33) $2\overline{)64}=$

34) $3\overline{)48}=$

35) $4\overline{)76}=$

36) $12\overline{)720}=$

37) $8\overline{)1,160}=$

38) $6\overline{)750}=$

39) $9\overline{)3,168}=$

40) $4\overline{)812}=$

41) $5\overline{)1,025}=$

42) $3\overline{)2,589}=$

Division with Remainders

Find the quotient with remainder.

1) $6\overline{)38}$

2) $5\overline{)29}$

3) $8\overline{)67}$

4) $3\overline{)10}$

5) $7\overline{)53}$

6) $3\overline{)17}$

7) $12\overline{)146}$

8) $21\overline{)444}$

9) $4\overline{)22}$

10) $5\overline{)48}$

11) $10\overline{)71}$

12) $8\overline{)24}$

13) $7\overline{)51}$

14) $9\overline{)84}$

15) $6\overline{)40}$

16) $12\overline{)126}$

17) $15\overline{)228}$

18) $11\overline{)177}$

19) $13\overline{)38}$

20) $3\overline{)20}$

21) $13\overline{)222}$

22) $5\overline{)46}$

23) $4\overline{)9}$

24) $9\overline{)1450}$

25) $96\overline{)194}$

26) $36\overline{)183}$

27) $38\overline{)230}$

28) $146\overline{)443}$

29) $42\overline{)1,766}$

30) $92\overline{)554}$

31) $210\overline{)632}$

32) $135\overline{)810}$

33) $6\overline{)79}$

34) $13\overline{)161}$

35) $126\overline{)885}$

36) $85\overline{)853}$

37) $125\overline{)1502}$

38) $11\overline{)4,832}$

39) $8\overline{)2,691}$

40) $7\overline{)953}$

41) $3\overline{)2,265}$

42) $4\overline{)6,744}$

Dividing Tens and Hundreds

Find answers.

1) $2000 \div 200$

2) $1600 \div 20$

3) $900 \div 100$

4) $3,200 \div 800$

5) $4,800 \div 800$

6) $900 \div 300$

7) $2,400 \div 800$

8) $4,500 \div 900$

9) $6,800 \div 200$

10) $10,000 \div 200$

11) $8,100 \div 300$

12) $8,000 \div 500$

13) $1,200 \div 200$

14) $6,600 \div 600$

15) $7,200 \div 600$

16) $1,800 \div 200$

17) $27,000 \div 900$

18) $9,900 \div 300$

19) $7,200 \div 100$

20) $9,000 \div 120$

21) $9,000 \div 3,000$

22) $16,000 \div 40$

23) $210 \div 30$

24) $560 \div 70$

Division Missing Number

Find each missing number.

1) $16 \div \underline{} = 2$

2) $\underline{} \div 8 = 6$

3) $18 \div \underline{} = 2$

4) $\underline{} \div 5 = 9$

5) $32 \div \underline{} = 4$

6) $\underline{} \div 9 = 8$

7) $40 \div \underline{} = 5$

8) $240 \div 16 = \underline{}$

9) $99 \div \underline{} = 9$

10) $80 \div 10 = \underline{}$

11) $24 \div \underline{} = 3$

12) $42 \div \underline{} = 6$

13) $\underline{} \div 8 = 7$

14) $120 \div 40 = \underline{}$

15) $18 \div \underline{} = 1$

16) $60 \div \underline{} = 6$

17) $\underline{} \div 14 = 9$

18) $\underline{} \div 11 = 13$

19) $70 \div \underline{} = 7$

20) $\underline{} \div 10 = 3$

21) $49 \div 7 = \underline{}$

22) $100 \div 10 = \underline{}$

23) $14 \div 14 = \underline{}$

24) $625 \div \underline{} = 25$

25) Linda planted 180 seeds. She wants to put them in equal numbers on 6 rows.

How many seed can she put on a row? _____ seeds

26) If dividend is 144 and the quotient is 16, then what is the divisor? _____

Answer key Chapter 3

Multiplication whole numbers

1) 636
2) 460
3) 204
4) 630
5) 576
6) 945

7) 156
8) 840
9) 980
10) 2,256
11) 1,452
12) 720

13) 7,935
14) 21,750
15) 9,100
16) 119
17) 432

Multiply Tens and Hundreds300

1) 2,800
2) 270
3) 560
4) 2,400
5) 6,300
6) 12,000
7) 9,000
8) 48,000
9) 4,900

10) 80
11) 800
12) 300
13) 80
14) 12,00
15) 9
16) 40
17) 90
18) 7

19) 40
20) 50
21) 60
22) 100
23) 7
24) 500
25) 90
26) 60
27) 60

Estimate products

1) 800
2) 200
3) 400
4) 200
5) 1,400
6) 2,800
7) 3,600
8) 800

9) 200
10) 2,100
11) 800
12) 500
13) 1,800
14) 4,800
15) 4,000
16) 1,800

17) 1,200
18) 1,500
19) 1,200
20) 2,000
21) 4,000
22) 3,000
23) 500
24) 3,600

Multiplication Missing Numbers

1) 4
2) 3
3) 8

4) 4
5) 6
6) 9

7) 8
8) 30
9) 84

10) 180	16) 8	22) 150
11) 88	17) 9	23) 32
12) 207	18) 125	24) 12
13) 6	19) 13	25) 168
14) 40	20) 12	
15) 4	21) 240	

Long Division by One Digit

1) 20	15) 14	29) 33
2) 8	16) 15	30) 20
3) 13	17) 18	31) 44
4) 8	18) 3	32) 80
5) 12	19) 15	33) 32
6) 6	20) 17	34) 16
7) 6	21) 19	35) 19
8) 3	22) 18	36) 60
9) 52	23) 117	37) 145
10) 6	24) 42	38) 125
11) 9	25) 135	39) 352
12) 5	26) 80	40) 203
13) 7	27) 27	41) 205
14) 10	28) 14	42) 863

Division with Remainders

1) 6 R2	10) 9 R3	19) 2 R12
2) 5 R4	11) 7 R1	20) 6 R2
3) 8 R3	12) 3 R0	21) 17 R1
4) 3 R1	13) 7 R2	22) 9 R1
5) 7 R4	14) 9 R3	23) 2 R1
6) 5 R2	15) 6 R4	24) 161 R1
7) 12 R2	16) 10 R6	25) 2 R2
8) 21 R3	17) 15 R3	26) 5 R3
9) 4 R2	18) 16 R1	27) 6 R2

28) 3 R5	33) 13 R1	38) 439 R3
29) 42 R2	34) 12 R5	39) 336 R3
30) 6 R2	35) 7 R3	40) R8
31) 3 R2	36) 10 R3	41) 753 R6
32) 6 R0	37) 12 R2	42) 1,685 R4

Dividing Tens and Hundreds

1) 10	9) 34	17) 30
2) 80	10) 50	18) 33
3) 9	11) 27	19) 75
4) 4	12) 16	20) 75
5) 6	13) 6	21) 3
6) 3	14) 11	22) 400
7) 3	15) 12	23) 7
8) 5	16) 9	24) 8

Division Missing Number

1) 8	10) 8	19) 10
2) 48	11) 8	20) 30
3) 9	12) 7	21) 7
4) 45	13) 56	22) 10
5) 8	14) 3	23) 1
6) 72	15) 18	24) 25
7) 8	16) 10	25) 30
8) 15	17) 126	26) 9
9) 11	18) 143	

Chapter 4:

Number Theory

Factoring

Factor, write prime if prime.

1) 15

2) 72

3) 25

4) 42

5) 32

6) 66

7) 34

8) 20

9) 50

10) 35

11) 40

12) 30

13) 49

14) 54

15) 96

16) 108

17) 76

18) 90

19) 100

20) 85

21) 63

22) 24

23) 48

24) 115

25) 51

26) 93

27) 52

28) 105

Prime Factorization

Factor the following numbers to their prime factors.

1.

$$16$$
/ \
_____

2.

$$38$$
/ \
_____

3.

$$51$$
/ \
_____

4.

$$12$$
/ \
_____

5.

$$18$$
/ \
_____

6.

$$23$$
/ \
_____

7.

$$46$$
/ \
_____

8.

$$58$$
/ \
_____

9.

$$64$$
/ \
_____

10.

$$82$$
/ \
_____

11.

$$87$$
/ \
_____

12.

$$98$$
/ \
_____

Divisibility Rule

Apply the divisibility rules to find the factors of each number.

1) 20	2, 3, 4, 5, 6, 9, 10	13) 24	2, 3, 4, 5, 6, 9, 10
2) 84	2, 3, 4, 5, 6, 9, 10	14) 395	2, 3, 4, 5, 6, 9, 10
3) 252	2, 3, 4, 5, 6, 9, 10	15) 920	2, 3, 4, 5, 6, 9, 10
4) 64	2, 3, 4, 5, 6, 9, 10	16) 137	2, 3, 4, 5, 6, 9, 10
5) 220	2, 3, 4, 5, 6, 9, 10	17) 440	2, 3, 4, 5, 6, 9, 10
6) 465	2, 3, 4, 5, 6, 9, 10	18) 360	2, 3, 4, 5, 6, 9, 10
7) 75	2, 3, 4, 5, 6, 9, 10	19) 495	2, 3, 4, 5, 6, 9, 10
8) 120	2, 3, 4, 5, 6, 9, 10	20) 4,870	2, 3, 4, 5, 6, 9, 10
9) 1,125	2, 3, 4, 5, 6, 9, 10	21) 590	2, 3, 4, 5, 6, 9, 10
10) 88	2, 3, 4, 5, 6, 9, 10	22) 326	2, 3, 4, 5, 6, 9, 10
11) 454	2, 3, 4, 5, 6, 9, 10	23) 114	2, 3, 4, 5, 6, 9, 10
12) 155	2, 3, 4, 5, 6, 9, 10	24) 470	2, 3, 4, 5, 6, 9, 10

Great Common Factor (GCF)

Find the GCF of the numbers.

1) 8, 26

2) 16, 44

3) 28, 38

4) 10, 35

5) 18, 48

6) 36, 52

7) 40, 75

8) 50, 45

9) 52, 8

10) 55, 85

11) 74, 94

12) 65, 20

13) 90, 10

14) 12, 34

15) 58, 86

16) 40, 95

17) 14, 56

18) 70, 100, 30

19) 64, 108

20) 63, 91

21) 20, 15, 35

22) 6, 12, 36

23) 25, 35, 70

24) 41, 39

Least Common Multiple (LCM)

Find the LCM of each.

1) 6, 14

2) 12, 18

3) 4, 16, 12

4) 10, 8

5) 10, 2, 15

6) 35, 7

7) 14, 35, 21

8) 8, 7

9) 11, 22, 44

10) 42, 21

11) 24, 72

12) 100, 25

13) 10, 5, 20

14) 15, 60

15) 20, 4, 3

16) 21, 14

17) 34, 17

18) 16, 64

19) 20, 70

20) 9, 39

21) 13, 8

22) 7, 20

23) 45, 63

24) 27, 4

Distributive Property

Multiply using the distributive property.

1) $3(x + 3) =$ _____

2) $4(x + 11) =$ _____

3) $(x + 9)5 =$ _____

4) $7(x + 6) =$ _____

5) $8(x + 8) =$ _____

6) $10(x + 4) =$ _____

7) $9(x + 10) =$ _____

8) $4(x + 8) =$ _____

9) $11(x + 6) =$ _____

10) $(x + 7)6 =$ _____

11) $(x + 13)4 =$ _____

12) $3(x + 12) =$ _____

13) $2(9x - 4) =$ _____

14) $7(8x - 3) =$ _____

15) $8(9x - 5) =$ _____

16) $(4x - 2)3 =$ _____

17) $(7x - 2)7 =$ _____

18) $(2x - 3)12 =$ _____

19) $3(4x - 1) =$ _____

20) $(-2)(4x - 3) =$ _____

21) $(-5)(x - 9) =$ _____

22) $(-7)(3x - 1) =$ _____

23) $(5x + 2)(-9) =$ _____

24) $(x + 6)(-12) =$ _____

Answer key Chapter 4

Factoring

1) 1, 3, 5, 15

2) 1, 2, 3, 4, 6, 8, 9, 12, 18, 24, 36, 72

3) 1, 5, 25

4) 1, 2, 3, 6, 7, 14, 21, 42

5) 1, 2, 4, 8, 16, 32

6) 1, 2, 3, 6, 11, 22, 33, 66

7) 1, 2, 17, 34

8) 1, 2, 4, 5, 10, 20

9) 1, 2, 5, 10, 25, 50

10) 1, 5, 7, 35

11) 1, 2, 4, 5, 8, 10, 20,40

12) 1, 2, 3, 5, 6, 10, 15, 30

13) 1, 7, 49

14) 1, 2, 3, 6, 9, 18, 27, 54

15) 1, 2, 3, 4, 6, 8, 12, 16, 24, 32, 48, 96

16) 1, 2, 3, 4, 6, 9, 12, 18, 27, 36, 54, 108

17) 1, 2, 4, 19, 38, 76

18) 1, 2, 3, 5, 6, 9, 10, 15, 18, 30, 45, 90

19) 1, 2, 4, 5, 10, 20, 25, 50, 100

20) 1, 5, 17, 85

21) 1, 3, 7, 9, 21, 63

22) 1, 2, 3, 4, 6, 8, 12, 24

23) 1, 2, 3, 4, 6, 8, 12, 16, 24, 48

24) 1, 5, 23, 115

25) 1, 3, 17, 51

26) 1, 3, 31, 93

27) 1, 2, 4, 13, 26, 52

28) 1, 3, 5, 7, 15, 21, 35, 105

Prime Factorization

1) $2 \times 2 \times 2 \times 2$

2) 2×19

3) 3×17

4) $2 \times 2 \times 3$

5) $2 \times 3 \times 3$

6) 23 is a prime number

7) 2×23

8) 2×29

9) $2 \times 2 \times 2 \times 2 \times 2 \times 2$

10) 2×41

11) 3×29

12) $2 \times 7 \times 7$

Divisibility Rule

1) 20   <u>2</u>, 3, <u>4</u>, <u>5</u>, 6, 9, 10

2) 84   <u>2</u>, <u>3</u>, <u>4</u>, 5, <u>6</u>, 9, 10

3) 252   <u>2</u>, <u>3</u>, <u>4</u>, 5, <u>6</u>, <u>9</u>, 10

4) 64   <u>2</u>, 3, <u>4</u>, 5, 6, 9, 10

5) 220   <u>2</u>, 3, <u>4</u>, <u>5</u>, 6, 9, <u>10</u>

6) 465   2, <u>3</u>, 4, <u>5</u>, 6, 9, 10

7) 75   2, <u>3</u>, 4, <u>5</u>, 6, 9, 10

8) 120   <u>2</u>, <u>3</u>, <u>4</u>, <u>5</u>, <u>6</u>, 9, <u>10</u>

9) 1,125   2, <u>3</u>, 4, <u>5</u>, 6, <u>9</u>, 10

10) 88   <u>2</u>, 3, <u>4</u>, 5, 6, 9, 10

11) 454   <u>2</u>, 3, 4, 5, 6, 9, 10

12) 155   2, 3, 4, <u>5</u>, 6, 9, 10

13) 24   <u>2</u>, <u>3</u>, <u>4</u>, 5, <u>6</u>, 9, 10

14) 395   2, 3, 4, <u>5</u>, 6, 9, 10

15) 920   <u>2</u>, 3, <u>4</u>, <u>5</u>, 6, 9, <u>10</u>

16) 137   2, 3, 4, 5, 6, 9, 10

17) 440   <u>2</u>, 3, <u>4</u>, <u>5</u>, 6, 9, <u>10</u>

18) 360   <u>2</u>, <u>3</u>, <u>4</u>, <u>5</u>, <u>6</u>, <u>9</u>, <u>10</u>

19) 495   2, <u>3</u>, 4, <u>5</u>, 6, <u>9</u>, 10

20) 4,870   <u>2</u>, 3, 4, <u>5</u>, 6, 9, <u>10</u>

21) 590   <u>2</u>, 3, 4, <u>5</u>, 6, 9, <u>10</u>

22) 326   <u>2</u>, 3, 4, 5, 6, 9, 10

23) 114   <u>2</u>, <u>3</u>, 4, 5, <u>6</u>, 9, 10

24) 470   <u>2</u>, 3, 4, <u>5</u>, 6, 9, <u>10</u>

Great Common Factor (GCF)

1) 2

2) 4

3) 2

4) 5

5) 6

6) 4

7) 5

8) 5

9) 4

10) 5

11) 2

12) 5

13) 10

14) 2

15) 2

16) 5

17) 14

18) 10

19) 4

20) 7

21) 5

22) 6

23) 5

24) 1

Least Common Multiple (LCM)

1) 42

2) 36

3) 48

4) 40

5) 30

6) 35

7) 210

8) 56

9) 44

10) 42

11) 72

12) 100

13) 20

14) 60

15) 60

16) 42

17) 34

18) 64

19) 140

20) 117

21) 104

22) 140

23) 315

24) 108

Distributive Property

1) $3x + 9$

2) $4x + 44$

3) $5x + 45$

4) $7x + 42$

5) $8x + 64$

6) $10x + 40$

7) $9x + 90$

8) $4x + 32$

9) $11x + 66$

10) $6x + 42$

11) $4x + 52$

12) $3x + 36$

13) $18x - 8$

14) $56x - 21$

15) $72x - 40$

16) $12x - 6$

17) $49x - 14$

18) $24x - 36$

19) $12x - 3$

20) $-8x + 6$

21) $-5x + 45$

22) $-21x + 7$

23) $-45x - 18$

24) $-12x - 72$

Chapter 5: Patterns

Repeating Pattern

Circle the picture that comes next in each picture pattern.

1)

2)

3)

4)

5)

6)

7)

Growing Patterns

Draw the picture that comes next in each growing pattern.

1)

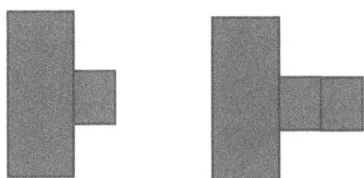

2)

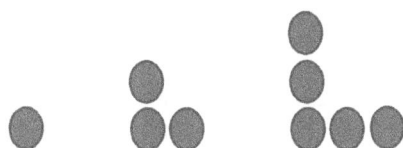

3)

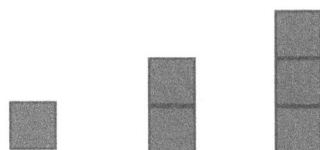

4)

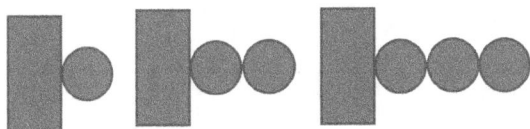

5)

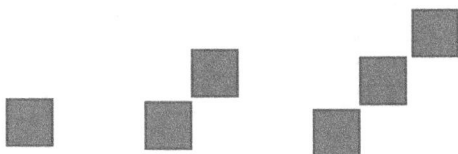

6)

Patterns: Numbers

Continue this pattern for four more numbers:

12)   1,500; 1,350; 1,200; 1,050; _____

13)   2,800; 2,600; 2,400; 2,200; _____

14)   3,500; 3,150; 2,800; 2,450; _____

15)   1,900; 1,780; 1,660; 1,540; _____

16)   3,200; 2,950; 2,700; 2,450; _____

17)   4,100; 3,800; 3,500; 3,200; _____

18)   5,400; 4,950; 4,500; 4,050; _____

19)   2,900; 2,725; 2,550; 2,375; _____

20)   1,950; 1,700; 1,450; 1,200; _____

21) 5,500; 4,900; 4,300; 3,700; _____

22) Write a list of five numbers that follows this pattern: Start at 100 and add 400 each time.

_____

Find a Rule

Complete the output.

1- **Rule: the output is $x + 25$**

Input	x	8	15	20	38	40
Output	y					

2- **Rule: the output is $x \times 18$**

Input	x	3	7	10	11	15
Output	y					

3- **Rule: the output is $x \div 7$**

Input	x	126	147	105	280	455
Output	y					

Find a rule to write an expression.

4- **Rule:** _____

Input	x	11	13	15	20
Output	y	55	65	75	100

5- **Rule:** _____

Input	x	10	28	32	46
Output	y	14	32	36	50

6- **Rule:** _____

Input	x	84	132	180	252
Output	y	14	22	30	42

Algebraic Thinking

Circle the number sentence that fits the problem. Then solve for x.

1) Mary had $42. Then she earned more money (x). Now she has $86.

   $42 + x = $86 OR $42 + $86 = x

   x = ____

2) Lisa had $35. Then she earned more money (x). Now she has $78.

   $35 + x = $78 OR $35 + $78 = x

   x = ____

3) Matthew had $37. Then he earned more money (x). Now he has $98.

   $37 + x = $98   OR $37 + $98 = x

   x = ____

4) Charlotte gave 19 of the cookies he had baked to a friend and now he has 45

   cookies left. 45 – 19 = x OR x – 19 = 45

   x = ____

5) Mia gave 32 of the cookies she had baked to a friend and now she has 55

   cookies left. 55 – 32 = x OR x – 32 = 55

   x = ____

6) Lucas gave 41 of the cookies he had baked to a friend and now he has 49

   cookies left. . 49 – 41 = x OR x – 41 = 49

   x = ____

Answers of Worksheets – Chapter 5

Repeating pattern

1)

2)

3)

4)

5)

6)

7)

Growing patterns

1)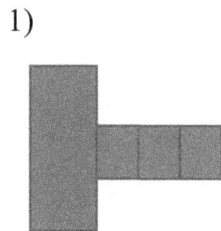

2)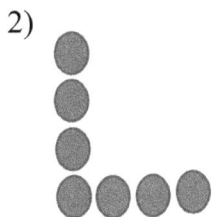

3)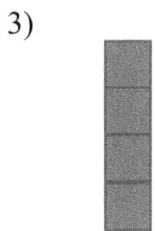

4)

5)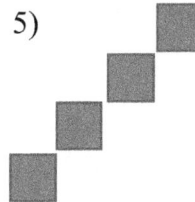

6)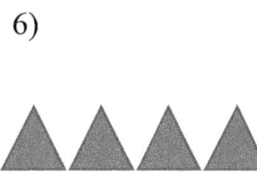

Pattern

12) 900; 750; 600; 450

13) 2,000; 1,800; 1,600; 1,400

14) 2,100; 1,750; 1,400; 1,050

15) 1,420; 1,300; 1,180; 1,060

16) 2,200; 1,950; 1,700; 1,450

17) 2,900; 2,600; 2,300; 2,000

18) 3,600; 3,150; 2,700; 2,250

19) 2,200; 2,025; 1,850; 1,675

20) 950; 700; 450; 200

21) 3,100; 2,500; 1,900; 1,300

22) 100; 500; 900; 1,300; 1,700

Find a Rule

1)

Input	x	8	15	20	38	40
Output	y	33	40	45	63	**65**

2)

Input	x	3	7	10	11	15
Output	y	54	**126**	180	198	270

3)

Input	x	126	147	105	280	455
Output	y	18	21	15	40	65

4) y = 5x

5) y = x + 4

6) y = x ÷ 6

Algebraic Thinking

1) $42 + x = $86; $x = 44$

2) $35 + x = $78; $x = 43$

3) $37 + x = $98; $x = 61$

4) $x - 19 = 45; x = 64$

5) $x - 32 = 55; x = 87$

6) $x - 41 = 49; x = 90$

Chapter 6: Fractions and Mix Numbers

Adding Fractions – Like Denominator

Find each sum.

1) $\dfrac{1}{3} + \dfrac{1}{3} =$

2) $\dfrac{3}{7} + \dfrac{1}{7} =$

3) $\dfrac{2}{9} + \dfrac{5}{9} =$

4) $\dfrac{7}{15} + \dfrac{1}{15} =$

5) $\dfrac{5}{23} + \dfrac{4}{23} =$

6) $\dfrac{8}{29} + \dfrac{7}{29} =$

7) $\dfrac{7}{19} + \dfrac{1}{19} =$

8) $\dfrac{5}{16} + \dfrac{1}{16} =$

9) $\dfrac{5}{31} + \dfrac{8}{31} =$

10) $\dfrac{5}{51} + \dfrac{8}{51} =$

11) $\dfrac{1}{17} + \dfrac{3}{17} =$

12) $\dfrac{2}{13} + \dfrac{4}{13} =$

13) $\dfrac{5}{41} + \dfrac{19}{41} =$

14) $\dfrac{2}{55} + \dfrac{9}{55} =$

15) $\dfrac{5}{21} + \dfrac{8}{21} =$

16) $\dfrac{10}{33} + \dfrac{2}{33} =$

17) $\dfrac{3}{11} + \dfrac{3}{11} =$

18) $\dfrac{24}{67} + \dfrac{1}{67} =$

19) $\dfrac{3}{19} + \dfrac{6}{19} =$

20) $\dfrac{20}{47} + \dfrac{13}{47} =$

Adding Fractions – Unlike Denominator

Add the fractions and simplify the answers.

1) $\frac{1}{2} + \frac{1}{6} =$

2) $\frac{2}{3} + \frac{3}{4} =$

3) $\frac{3}{5} + \frac{1}{2} =$

4) $\frac{7}{10} + \frac{1}{3} =$

5) $\frac{4}{15} + \frac{1}{5} =$

6) $\frac{3}{14} + \frac{2}{7} =$

7) $\frac{2}{7} + \frac{1}{3} =$

8) $\frac{1}{20} + \frac{1}{5} =$

9) $\frac{7}{12} + \frac{1}{6} =$

10) $\frac{1}{8} + \frac{3}{4} =$

11) $\frac{4}{21} + \frac{1}{3} =$

12) $\frac{5}{36} + \frac{2}{9} =$

13) $\frac{4}{35} + \frac{3}{7} =$

14) $\frac{2}{33} + \frac{1}{11} =$

15) $\frac{10}{27} + \frac{1}{3} =$

16) $\frac{7}{45} + \frac{4}{9} =$

17)

18) $\frac{1}{7} + \frac{2}{5} =$

19) $\frac{1}{2} + \frac{5}{22} =$

20) $\frac{3}{16} + \frac{1}{4} =$

21) $\frac{5}{16} + \frac{1}{24} =$

22) $\frac{2}{15} + \frac{3}{10} =$

23) $\frac{5}{42} + \frac{4}{21} =$

24) $\frac{1}{36} + \frac{5}{24} =$

Subtracting Fractions – Like Denominator

Find the difference.

1) $\dfrac{6}{5} - \dfrac{1}{5} =$

2) $\dfrac{7}{12} - \dfrac{4}{12} =$

3) $\dfrac{8}{13} - \dfrac{5}{13} =$

4) $\dfrac{19}{6} - \dfrac{7}{6} =$

5) $\dfrac{11}{21} - \dfrac{9}{21} =$

6) $\dfrac{15}{43} - \dfrac{7}{43} =$

7) $\dfrac{9}{23} - \dfrac{3}{23} =$

8) $\dfrac{18}{47} - \dfrac{15}{47} =$

9) $\dfrac{8}{20} - \dfrac{4}{20} =$

10) $\dfrac{34}{48} - \dfrac{17}{48} =$

11) $\dfrac{6}{7} - \dfrac{2}{7} =$

12) $\dfrac{36}{51} - \dfrac{28}{51} =$

13) $\dfrac{9}{11} - \dfrac{5}{11} =$

14) $\dfrac{29}{49} - \dfrac{14}{49} =$

15) $\dfrac{12}{17} - \dfrac{6}{17} =$

16) $\dfrac{17}{23} - \dfrac{11}{23} =$

17) $\dfrac{5}{7} - \dfrac{1}{7} =$

18) $\dfrac{12}{31} - \dfrac{8}{31} =$

19) $\dfrac{38}{61} - \dfrac{29}{61} =$

20) $\dfrac{31}{52} - \dfrac{20}{52} =$

21) $\dfrac{51}{63} - \dfrac{46}{63} =$

22) $\dfrac{55}{83} - \dfrac{25}{83} =$

23) $\dfrac{54}{77} - \dfrac{30}{77} =$

24) $\dfrac{49}{55} - \dfrac{37}{55} =$

Subtracting Fractions – Unlike Denominator

Solve each problem.

1) $\frac{1}{3} - \frac{1}{6} =$

2) $\frac{3}{5} - \frac{1}{9} =$

3) $\frac{1}{4} - \frac{1}{6} =$

4) $\frac{7}{9} - \frac{3}{10} =$

5) $\frac{11}{12} - \frac{5}{24} =$

6) $\frac{9}{16} - \frac{3}{20} =$

7) $\frac{19}{30} - \frac{1}{6} =$

8) $\frac{1}{2} - \frac{7}{15} =$

9) $\frac{3}{5} - \frac{2}{7} =$

10) $\frac{8}{9} - \frac{4}{11} =$

11) $\frac{6}{8} - \frac{7}{48} =$

12) $\frac{3}{4} - \frac{7}{10} =$

13) $\frac{4}{5} - \frac{8}{45} =$

14) $\frac{6}{7} - \frac{3}{10} =$

15) $\frac{11}{12} - \frac{13}{24} =$

16) $\frac{4}{9} - \frac{15}{63} =$

17) $\frac{5}{12} - \frac{5}{16} =$

18) $\frac{5}{8} - \frac{3}{10} =$

19) $\frac{5}{7} - \frac{3}{8} =$

20) $\frac{3}{4} - \frac{21}{44} =$

Converting Mix Numbers

Convert the following mixed numbers into improper fractions.

1) $2\frac{3}{5} =$

2) $3\frac{4}{5} =$

3) $5\frac{3}{4} =$

4) $3\frac{5}{8} =$

5) $6\frac{2}{5} =$

6) $7\frac{8}{9} =$

7) $2\frac{2}{15} =$

8) $3\frac{1}{13} =$

9) $2\frac{1}{12} =$

10) $5\frac{5}{6} =$

11) $7\frac{3}{5} =$

12) $3\frac{9}{10} =$

13) $6\frac{1}{3} =$

14) $5\frac{5}{6} =$

15) $8\frac{2}{5} =$

16) $4\frac{3}{8} =$

17) $3\frac{5}{9} =$

18) $2\frac{3}{14} =$

19) $7\frac{5}{6} =$

20) $5\frac{6}{7} =$

21) $4\frac{7}{8} =$

22) $3\frac{2}{9} =$

23) $2\frac{9}{11} =$

24) $10\frac{5}{7} =$

Converting improper Fractions

Convert the following improper fractions into mixed numbers

1) $\frac{55}{16} =$

2) $\frac{95}{34} =$

3) $\frac{39}{19} =$

4) $\frac{11}{3} =$

5) $\frac{61}{17} =$

6) $\frac{152}{41} =$

7) $\frac{125}{31} =$

8) $\frac{46}{7} =$

9) $\frac{43}{11} =$

10) $\frac{14}{3} =$

11) $\frac{29}{6} =$

12) $\frac{61}{15} =$

13) $\frac{56}{24} =$

14) $\frac{21}{9} =$

15) $\frac{115}{16} =$

16) $\frac{59}{6} =$

17) $\frac{144}{10} =$

18) $\frac{51}{13} =$

19) $\frac{36}{8} =$

20) $\frac{58}{6} =$

21) $\frac{9}{7} =$

22) $\frac{89}{11} =$

23) $\frac{102}{9} =$

24) $\frac{180}{19} =$

Adding Mix Numbers

Add the following fractions.

1) $3\frac{4}{9} + 2\frac{2}{9} =$

2) $3\frac{2}{5} + 2\frac{1}{5} =$

3) $1\frac{1}{7} + 2\frac{3}{7} =$

4) $4\frac{5}{6} + 2\frac{1}{3} =$

5) $1\frac{4}{15} + 2\frac{2}{5} =$

6) $4\frac{1}{3} + 1\frac{3}{4} =$

7) $3\frac{7}{9} + 3\frac{1}{6} =$

8) $3\frac{5}{8} + 2\frac{1}{2} =$

9) $3\frac{3}{4} + 2\frac{1}{4} =$

10) $1\frac{4}{11} + 2\frac{3}{11} =$

11) $4\frac{1}{2} + 2\frac{2}{5} =$

12) $5\frac{1}{4} + 2\frac{5}{6} =$

13) $6\frac{2}{7} + 2\frac{5}{7} =$

14) $3\frac{7}{8} + 2\frac{3}{16} =$

15) $3\frac{3}{4} + 3\frac{2}{9} =$

16) $4\frac{3}{5} + 2\frac{1}{6} =$

17) $7\frac{1}{2} + 6\frac{3}{5} =$

18) $6\frac{2}{3} + 1\frac{5}{12} =$

19) $2\frac{1}{9} + 7\frac{2}{3} =$

20) $4\frac{1}{6} + 2\frac{5}{9} =$

21) $5\frac{2}{3} + 6\frac{3}{4} =$

22) $7\frac{1}{8} + 1\frac{7}{24} =$

23) $5\frac{3}{7} + 4\frac{1}{8} =$

24) $8\frac{1}{3} + 4\frac{3}{4} =$

Subtracting Mix Numbers

Subtract the following fractions.

1) $8\frac{1}{4} - 7\frac{1}{4} =$

2) $5\frac{5}{6} - 5\frac{1}{6} =$

3) $9\frac{7}{11} - 8\frac{3}{11} =$

4) $5\frac{1}{2} - 2\frac{1}{6} =$

5) $4\frac{1}{4} - 1\frac{1}{8} =$

6) $9\frac{1}{3} - 5\frac{3}{7} =$

7) $5\frac{7}{9} - 2\frac{2}{9} =$

8) $8\frac{15}{17} - 5\frac{11}{17} =$

9) $9\frac{11}{14} - 3\frac{5}{14} =$

10) $7\frac{9}{10} - 6\frac{7}{10} =$

11) $8\frac{3}{4} - 5\frac{1}{12} =$

12) $6\frac{7}{8} - 3\frac{1}{8} =$

13) $7\frac{12}{35} - 3\frac{3}{7} =$

14) $6\frac{1}{3} - 4\frac{1}{9} =$

15) $10\frac{6}{7} - 7\frac{3}{7} =$

16) $9\frac{2}{3} - 3\frac{1}{3} =$

17) $5\frac{4}{11} - 3\frac{2}{11} =$

18) $7\frac{3}{10} - 5\frac{1}{5} =$

19) $8\frac{5}{6} - 5\frac{1}{12} =$

20) $3\frac{3}{4} - 3\frac{7}{16} =$

21) $8\frac{8}{13} - 3\frac{1}{3} =$

22) $6\frac{5}{6} - 4\frac{7}{30} =$

23) $5\frac{6}{7} - 4\frac{4}{11} =$

24) $6\frac{10}{19} - 2\frac{9}{19} =$

Simplify Fractions

Reduce these fractions to lowest terms

1) $\dfrac{18}{12} =$

2) $\dfrac{22}{33} =$

3) $\dfrac{32}{40} =$

4) $\dfrac{27}{36} =$

5) $\dfrac{8}{24} =$

6) $\dfrac{15}{35} =$

7) $\dfrac{20}{35} =$

8) $\dfrac{56}{70} =$

9) $\dfrac{9}{81} =$

10) $\dfrac{40}{16} =$

11) $\dfrac{54}{72} =$

12) $\dfrac{40}{120} =$

13) $\dfrac{12}{20} =$

14) $\dfrac{7}{28} =$

15) $\dfrac{14}{49} =$

16) $\dfrac{58}{87} =$

17) $\dfrac{72}{27} =$

18) $\dfrac{48}{180} =$

19) $\dfrac{24}{64} =$

20) $\dfrac{48}{42} =$

21) $\dfrac{120}{240} =$

22) $\dfrac{54}{279} =$

23) $\dfrac{340}{68} =$

24) $\dfrac{150}{600} =$

Multiplying Fractions

Find the product.

1) $\frac{2}{3} \times \frac{5}{7} =$

2) $\frac{3}{11} \times \frac{4}{9} =$

3) $\frac{7}{24} \times \frac{3}{14} =$

4) $\frac{7}{16} \times \frac{24}{35} =$

5) $\frac{15}{21} \times \frac{3}{5} =$

6) $\frac{18}{20} \times \frac{4}{9} =$

7) $\frac{6}{7} \times \frac{7}{9} =$

8) $\frac{54}{79} \times 0 =$

9) $\frac{2}{6} \times \frac{12}{14} =$

10) $\frac{24}{14} \times \frac{7}{8} =$

11) $\frac{38}{36} \times \frac{18}{19} =$

12) $\frac{8}{10} \times \frac{5}{64} =$

13) $\frac{15}{4} \times \frac{16}{9} =$

14) $\frac{25}{8} \times \frac{4}{10} =$

15) $\frac{14}{63} \times \frac{9}{7} =$

16) $\frac{12}{20} \times 4 =$

17) $\frac{7}{33} \times \frac{66}{21} =$

18) $\frac{5}{16} \times \frac{8}{10} =$

19) $\frac{9}{10} \times \frac{4}{27} =$

20) $\frac{6}{42} \times \frac{7}{12} =$

21) $\frac{8}{19} \times \frac{1}{16} =$

22) $\frac{10}{7} \times \frac{4}{80} =$

23) $\frac{9}{12} \times \frac{4}{54} =$

24) $\frac{60}{400} \times \frac{200}{600} =$

Multiplying Mixed Number

Multiply. Reduce to lowest terms.

1) $3\frac{1}{4} \times 3\frac{1}{5} =$

12) $2\frac{1}{3} \times 1\frac{2}{9} =$

2) $2\frac{2}{7} \times 1\frac{1}{8} =$

13) $3\frac{1}{4} \times 2\frac{2}{5} =$

3) $1\frac{1}{4} \times 2\frac{3}{5} =$

14) $4\frac{1}{9} \times 3\frac{1}{3} =$

4) $2\frac{2}{9} \times 1\frac{1}{10} =$

15) $3\frac{1}{5} \times 2\frac{1}{7} =$

5) $3\frac{3}{5} \times 2\frac{1}{5} =$

16) $4\frac{1}{2} \times 2\frac{2}{5} =$

6) $2\frac{3}{4} \times 2\frac{2}{3} =$

17) $1\frac{1}{4} \times 2\frac{4}{5} =$

7) $4\frac{1}{2} \times 1\frac{1}{9} =$

18) $3\frac{1}{3} \times 1\frac{1}{4} =$

8) $2\frac{4}{5} \times 4\frac{1}{7} =$

19) $4\frac{4}{5} \times 1\frac{5}{7} =$

9) $3\frac{1}{4} \times 2\frac{1}{3} =$

20) $6\frac{3}{4} \times 1\frac{1}{3} =$

10) $1\frac{1}{5} \times 5\frac{1}{6} =$

21) $5\frac{1}{2} \times 3\frac{1}{5} =$

11) $5\frac{1}{3} \times 2\frac{1}{8} =$

22) $4\frac{2}{3} \times 6\frac{1}{2} =$

Comparing Fractions

Compare the fractions, and write >, < or =

1) $\frac{13}{2}$ _____ $\frac{19}{10}$

2) $\frac{15}{4}$ _____ $\frac{5}{7}$

3) $\frac{5}{8}$ _____ $\frac{4}{5}$

4) $\frac{11}{3}$ _____ $\frac{12}{8}$

5) $\frac{2}{9}$ _____ $\frac{4}{7}$

6) $\frac{13}{5}$ _____ $\frac{17}{4}$

7) $\frac{14}{9}$ _____ $\frac{8}{11}$

8) $\frac{15}{13}$ _____ $\frac{21}{8}$

9) $5\frac{1}{10}$ _____ $8\frac{1}{15}$

10) $9\frac{1}{12}$ _____ $7\frac{1}{9}$

11) $4\frac{1}{4}$ _____ $4\frac{1}{7}$

12) $8\frac{6}{7}$ _____ $8\frac{3}{8}$

13) $1\frac{5}{9}$ _____ $4\frac{2}{3}$

14) $\frac{1}{24}$ _____ $\frac{2}{13}$

15) $\frac{42}{23}$ _____ $\frac{29}{72}$

16) $\frac{14}{200}$ _____ $\frac{8}{81}$

17) $19\frac{1}{3}$ _____ $19\frac{1}{7}$

18) $\frac{1}{8}$ _____ $\frac{1}{12}$

19) $\frac{1}{11}$ _____ $\frac{1}{15}$

20) $\frac{1}{15}$ _____ $\frac{7}{12}$

21) $\frac{10}{33}$ _____ $\frac{8}{59}$

22) $\frac{6}{7}$ _____ $\frac{3}{8}$

23) $6\frac{2}{5}$ _____ $4\frac{12}{5}$

24) $2\frac{12}{5}$ _____ $3\frac{4}{5}$

Answer key Chapter 6

Adding Fractions – Like Denominator

1) $\frac{2}{3}$

2) $\frac{4}{7}$

3) $\frac{7}{9}$

4) $\frac{8}{15}$

5) $\frac{9}{23}$

6) $\frac{15}{29}$

7) $\frac{8}{19}$

8) $\frac{3}{8}$

9) $\frac{13}{31}$

10) $\frac{13}{51}$

11) $\frac{4}{17}$

12) $\frac{6}{13}$

13) $\frac{24}{41}$

14) $\frac{1}{5}$

15) $\frac{13}{21}$

16) $\frac{4}{11}$

17) $\frac{6}{11}$

18) $\frac{25}{67}$

19) $\frac{9}{19}$

20) $\frac{33}{47}$

Adding Fractions – Unlike Denominator

1) $\frac{2}{3}$

2) $\frac{17}{12}$

3) $\frac{11}{10}$

4) $\frac{31}{30}$

5) $\frac{7}{15}$

6) $\frac{1}{2}$

7) $\frac{13}{21}$

8) $\frac{1}{4}$

9) $\frac{3}{4}$

10) $\frac{7}{8}$

11) $\frac{11}{21}$

12) $\frac{13}{36}$

13) $\frac{19}{35}$

14) $\frac{5}{33}$

15) $\frac{19}{27}$

16) $\frac{3}{5}$

17) $\frac{19}{45}$

18) $\frac{19}{35}$

19) $\frac{8}{11}$

20) $\frac{7}{16}$

21) $\frac{17}{48}$

22) $\frac{13}{30}$

23) $\frac{13}{42}$

24) $\frac{17}{72}$

Subtracting Fractions – Like Denominator

1) 1

2) $\frac{1}{4}$

3) $\frac{3}{13}$

4) 2

5) $\frac{2}{21}$

6) $\frac{8}{43}$

7) $\frac{6}{23}$

8) $\frac{3}{47}$

9) $\frac{1}{5}$

10) $\frac{17}{48}$

11) $\frac{4}{7}$

12) $\frac{8}{51}$

13) $\frac{4}{11}$

14) $\frac{15}{49}$

15) $\frac{6}{17}$

16) $\frac{6}{23}$

17) $\frac{4}{7}$

18) $\frac{4}{31}$

19) $\frac{9}{61}$

20) $\frac{11}{52}$

21) $\frac{5}{63}$

22) $\frac{30}{83}$

23) $\frac{24}{77}$

24) $\frac{12}{55}$

Subtracting Fractions – Unlike Denominator

1) $\frac{1}{6}$

2) $\frac{22}{45}$

3) $\frac{1}{12}$

4) $\frac{43}{90}$

5) $\frac{17}{24}$

6) $\frac{33}{80}$

7) $\frac{7}{15}$

8) $\frac{1}{30}$

9) $\frac{11}{35}$

10) $\frac{52}{99}$

11) $\frac{29}{48}$

12) $\frac{1}{20}$

13) $\frac{28}{45}$

14) $\frac{39}{70}$

15) $\frac{3}{8}$

16) $\frac{13}{63}$

17) $\frac{5}{48}$

18) $\frac{13}{40}$

19) $\frac{19}{56}$

20) $\frac{3}{11}$

Converting Mix Numbers

1) $\frac{13}{5}$

2) $\frac{19}{5}$

3) $\frac{23}{4}$

4) $\frac{29}{8}$

5) $\frac{32}{5}$

6) $\frac{71}{9}$

7) $\frac{32}{15}$

8) $\frac{40}{13}$

9) $\frac{25}{12}$

10) $\frac{35}{6}$

11) $\frac{38}{5}$

12) $\frac{39}{10}$

13) $\frac{19}{3}$

14) $\frac{35}{6}$

15) $\frac{42}{5}$

16) $\frac{35}{8}$

17) $\frac{32}{9}$

18) $\frac{31}{14}$

19) $\frac{47}{6}$

20) $\frac{41}{7}$

21) $\frac{39}{8}$

22) $\frac{29}{9}$

23) $\frac{31}{11}$

24) $\frac{75}{7}$

Converting improper Fractions

1) $3\frac{7}{16}$

2) $2\frac{27}{34}$

3) $2\frac{1}{19}$

4) $3\frac{2}{3}$

5) $3\frac{10}{17}$

6) $3\frac{29}{41}$

7) $4\frac{1}{31}$

8) $6\frac{4}{7}$

9) $3\frac{10}{11}$

10) $4\frac{2}{3}$

11) $4\frac{5}{6}$

12) $4\frac{1}{15}$

13) $2\frac{1}{3}$

14) $2\frac{1}{3}$

15) $7\frac{3}{16}$

16) $9\frac{5}{6}$

17) $14\frac{2}{5}$

18) $3\frac{12}{13}$

19) $4\frac{1}{2}$

20) $9\frac{2}{3}$

21) $1\frac{2}{7}$

22) $8\frac{1}{11}$

23) $11\frac{1}{3}$

24) $9\frac{9}{19}$

Adding Mix Numbers

1) $5\frac{2}{3}$

2) $5\frac{3}{5}$

3) $3\frac{4}{7}$

4) $7\frac{1}{6}$

5) $3\frac{2}{3}$

6) $6\frac{1}{12}$

7) $6\frac{17}{18}$

8) $6\frac{1}{8}$

9) 6

10) $3\frac{7}{11}$

11) $6\frac{9}{10}$

12) $8\frac{1}{12}$

13) 9

14) $6\frac{1}{16}$

15) $6\frac{35}{36}$

16) $6\frac{23}{30}$

17) $14\frac{1}{10}$

18) $8\frac{1}{12}$

19) $9\frac{7}{9}$

20) $6\frac{13}{18}$

21) $12\frac{5}{12}$

22) $8\frac{5}{12}$

23) $9\frac{31}{56}$

24) $13\frac{1}{12}$

Subtracting Mix Numbers

1) 1

2) $\frac{2}{3}$

3) $1\frac{4}{11}$

4) $3\frac{1}{3}$

5) $3\frac{1}{8}$

6) $3\frac{19}{21}$

7) $3\frac{5}{9}$

8) $3\frac{4}{17}$

9) $6\frac{3}{7}$

10) $1\frac{1}{5}$

11) $3\frac{2}{3}$

12) $3\frac{3}{4}$

13) $3\frac{32}{35}$

14) $2\frac{2}{9}$

15) $3\frac{3}{7}$

16) $6\frac{1}{3}$

17) $2\frac{2}{11}$

18) $2\frac{1}{10}$

19) $3\frac{3}{4}$

20) $\frac{5}{16}$

21) $5\frac{11}{39}$

22) $2\frac{3}{5}$

23) $1\frac{38}{77}$

24) $4\frac{1}{19}$

Simplify Fractions

1) $\frac{3}{2}$

2) $\frac{2}{3}$

3) $\frac{4}{5}$

4) $\frac{3}{4}$

5) $\frac{1}{3}$

6) $\frac{3}{7}$

7) $\frac{4}{7}$

8) $\frac{4}{5}$

9) $\frac{1}{9}$

10) $\frac{5}{2}$

11) $\frac{3}{4}$

12) $\frac{1}{3}$

13) $\frac{3}{5}$

14) $\frac{1}{4}$

15) $\frac{2}{7}$

16) $\frac{2}{3}$

17) $\frac{8}{3}$

18) $\frac{4}{15}$

19) $\frac{3}{8}$

20) $\frac{8}{7}$

21) $\frac{1}{2}$

22) $\frac{6}{31}$

23) 5

24) $\frac{1}{4}$

Multiplying Fractions

1) $\frac{10}{21}$

2) $\frac{4}{33}$

3) $\frac{1}{16}$

4) $\frac{3}{10}$

5) $\frac{3}{7}$

6) $\frac{2}{5}$

7) $\frac{2}{3}$

8) 0

9) $\frac{2}{7}$

10) $\frac{3}{2}$

11) 1

12) $\frac{1}{16}$

13) $\frac{20}{3}$

14) $\frac{5}{4}$

15) $\frac{2}{7}$

16) $\frac{12}{5}$

17) $\frac{2}{3}$

18) $\frac{1}{4}$

19) $\frac{2}{15}$

20) $\frac{1}{12}$

21) $\frac{1}{38}$

22) $\frac{1}{14}$

23) $\frac{1}{18}$

24) $\frac{1}{20}$

Multiplying Mixed Number

1) $10\frac{2}{5}$

2) $2\frac{4}{7}$

3) $3\frac{1}{4}$

4) $2\frac{4}{9}$

5) $7\frac{23}{25}$

6) $7\frac{1}{3}$

7) 5

8) $11\frac{3}{5}$

9) $7\frac{7}{12}$

10) $6\frac{1}{5}$

11) $11\frac{1}{3}$

12) $2\frac{23}{27}$

13) $7\frac{4}{5}$

14) $13\frac{19}{27}$

15) $6\frac{6}{7}$

16) $10\frac{4}{5}$

17) $3\frac{1}{2}$

18) $4\frac{1}{6}$

19) $8\frac{8}{35}$

20) 9

21) $17\frac{3}{5}$

22) $30\frac{1}{3}$

Comparing Fractions

1) >	7) >	13) <	19) >
2) >	8) <	14) <	20) <
3) <	9) <	15) >	21) >
4) >	10) >	16) <	22) >
5) <	11) >	17) >	23) =
6) <	12) >	18) >	24) >

Chapter 7: Decimal

Graph Decimals

Write the decimals indicated by the arrows.

1)

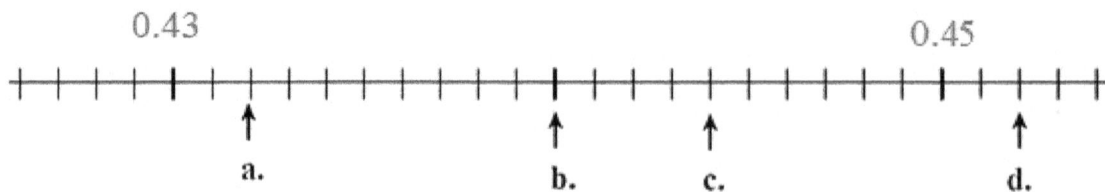

a. _____ b. _____ c. _____ d. _____

2)

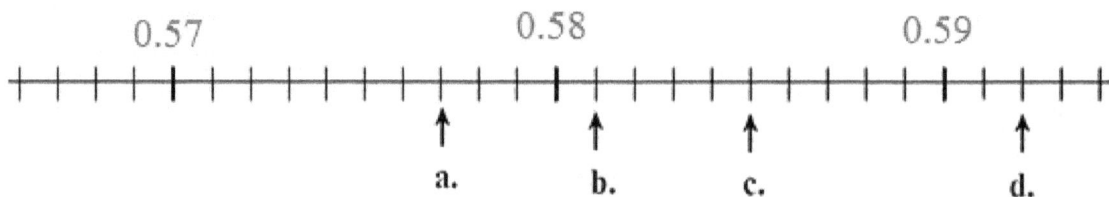

a. _____ b. _____ c. _____ d. _____

3)

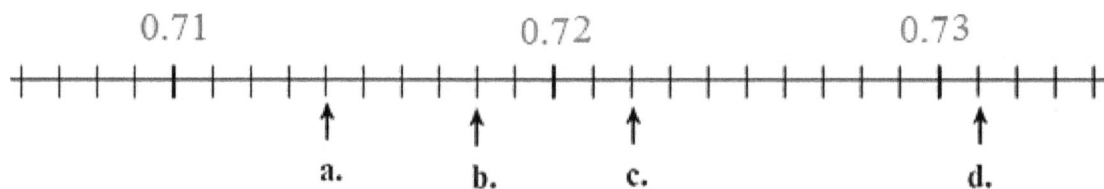

a. _____ b. _____ c. _____ d. _____

4)

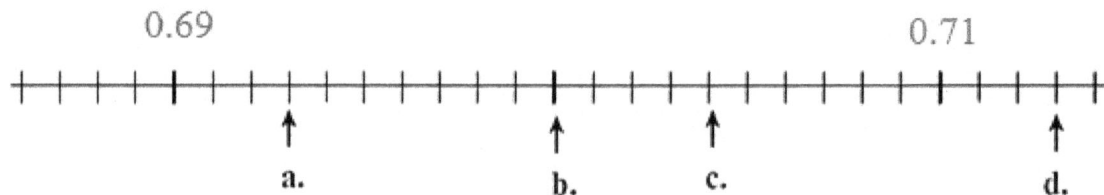

a. _____ b. _____ c. _____ d. _____

Round Decimals

Round each number to the correct place value

1) 0.9<u>2</u> =

2) 4.<u>2</u>1 =

3) 8.<u>8</u>33 =

4) 0.<u>5</u>69 =

5) <u>7</u>.779 =

6) 0.0<u>4</u>9 =

7) 9.<u>3</u>7 =

8) 25.3<u>3</u>1 =

9) 4.6<u>2</u>9 =

10) 10.<u>4</u>71 =

11) 3.<u>2</u>6 =

12) 4.<u>2</u>29 =

13) 6.<u>2</u>18 =

14) 9.1<u>6</u>24 =

15) 3<u>6</u>.29 =

16) 4<u>7</u>.68 =

17) 6<u>3</u>2.785 =

18) 577.<u>8</u>29 =

19) 24.5<u>7</u>9 =

20) 8<u>3</u>.91 =

21) 5.4<u>1</u>35 =

22) 86.<u>2</u>76 =

23) 354.<u>3</u>39 =

24) 0.9<u>2</u>66 =

25) 0.00<u>7</u>4 =

26) 8.0<u>4</u>49 =

27) 41.6<u>5</u>96 =

28) 19.0<u>8</u>27 =

Decimals Addition

Add the following.

1) 32.19
   + 31.16
   _____

2) 0.54
   + 0.31
   _____

3) 21.42
   + 12.57
   _____

4) 57.189
   + 7.221
   _____

5) 22.740
   + 8.37
   _____

6) 6.712
   + 4.105
   _____

7) 78.46
   + 18.54
   _____

8) 66.24
   + 20.36
   _____

9) 39.11
   + 15.09
   _____

10) 4.88
    + 19.45
    _____

11) 16.254
    + 34.227
    _____

12) 36.66
    + 4.82
    _____

13) 42.45
    + 9.37
    _____

14) 124.41
    + 5.71
    _____

Decimals Subtraction

Subtract the following

1)
$$\begin{array}{r} 8.19 \\ -\ 2.46 \\ \hline \end{array}$$

8)
$$\begin{array}{r} 45.65 \\ -\ 19.57 \\ \hline \end{array}$$

2)
$$\begin{array}{r} 46.25 \\ -\ 16.18 \\ \hline \end{array}$$

9)
$$\begin{array}{r} 65.42 \\ -\ 47.71 \\ \hline \end{array}$$

3)
$$\begin{array}{r} 0.89 \\ -\ 0.5 \\ \hline \end{array}$$

10)
$$\begin{array}{r} 8.652 \\ -\ 0.553 \\ \hline \end{array}$$

4)
$$\begin{array}{r} 24.354 \\ -7.8 \\ \hline \end{array}$$

11)
$$\begin{array}{r} 48.61 \\ -\ 29.52 \\ \hline \end{array}$$

5)
$$\begin{array}{r} 0.789 \\ -\ 0.06 \\ \hline \end{array}$$

12)
$$\begin{array}{r} 16.399 \\ -\ 5.389 \\ \hline \end{array}$$

6)
$$\begin{array}{r} 63.25 \\ -\ 42.52 \\ \hline \end{array}$$

13)
$$\begin{array}{r} 35.251 \\ -\ 9.169 \\ \hline \end{array}$$

7)
$$\begin{array}{r} 157.75 \\ -\ 94.87 \\ \hline \end{array}$$

14)
$$\begin{array}{r} 148.42 \\ -\ 11.78 \\ \hline \end{array}$$

Decimals Multiplication

Solve.

1) $\begin{array}{r} 2.4 \\ \times 1.9 \\ \hline \end{array}$

2) $\begin{array}{r} 3.6 \\ \times\ 5.2 \\ \hline \end{array}$

3) $\begin{array}{r} 4.02 \\ \times 2.04 \\ \hline \end{array}$

4) $\begin{array}{r} 45.9 \\ \times\ 10 \\ \hline \end{array}$

5) $\begin{array}{r} 49.8 \\ \times\ 100 \\ \hline \end{array}$

6) $\begin{array}{r} 41.56 \\ \times 4.2 \\ \hline \end{array}$

7) $\begin{array}{r} 24.51 \\ \times 10.2 \\ \hline \end{array}$

8) $\begin{array}{r} 1.89 \\ \times 7.35 \\ \hline \end{array}$

9) $\begin{array}{r} 14.05 \\ \times\ 0.09 \\ \hline \end{array}$

10) $\begin{array}{r} 51.03 \\ \times\ 4.04 \\ \hline \end{array}$

11) $\begin{array}{r} 14.56 \\ \times\ 12.4 \\ \hline \end{array}$

12) $\begin{array}{r} 9.56 \\ \times 0.04 \\ \hline \end{array}$

13) $\begin{array}{r} 7.5 \\ \times 0.11 \\ \hline \end{array}$

14) $\begin{array}{r} 23.1 \\ \times 4.02 \\ \hline \end{array}$

Decimal Division

Dividing Decimals.

1) $7 \div 1,000 =$

2) $3 \div 10,000 =$

3) $2.8 \div 10 =$

4) $0.08 \div 100 =$

5) $5 \div 25 =$

6) $4 \div 48 =$

7) $8 \div 40 =$

8) $7 \div 140 =$

9) $9 \div 10,000 =$

10) $0.6 \div 0.54 =$

11) $0.4 \div 0.004 =$

12) $0.3 \div 0.15 =$

13) $0.7 \div 0.56 =$

14) $0.6 \div 0.0006 =$

15) $4.9 \div 100 =$

16) $7.3 \div 100 =$

17) $8.5 \div 10 =$

18) $16.2 \div 4.4 =$

19) $36.9 \div 3.3 =$

20) $0.8 \div 0.08 =$

21) $8.04 \div 4.2 =$

22) $0.09 \div 0.30 =$

23) $0.8 \div 6.4 =$

24) $0.07 \div 42 =$

25) $6.28 \div 0.6 =$

26) $0.026 \div 13 =$

Comparing Decimals

Write the Correct Comparison Symbol (>, < or =)

1) 1.98 _____ 3.12

2) 0.6 _____ 0.549

3) 19.01 _____ 19.010

4) 5.05 _____ 5.50

5) 0.811 _____ 0.81

6) 0.658 _____ 0.865

7) 5.46 _____ 5.391

8) 7.021 _____ 7.035

9) 56.321 _____ 56.123

10) 4.69 _____ 4.069

11) 3.55 _____ 3.555

12) 0.08 _____ 0.12

13) 2.405 _____ 2.45

14) 7.53 _____ 7.35

15) 0.11 _____ 0.011

16) 86.09 _____ 86.090

17) 0.190 _____ 0.21

18) 53.98 _____ 54.07

19) 0.072 _____ 0.720

20) 43.2 _____ 34.9

21) 17.99 _____ 20.19

22) 0.087 _____ 0.0807

23) 3.059 _____ 0.3059

24) 8.2 _____ 0.825

25) 8.77 _____ 0.877

26) 9.56 _____ 9.5600

27) 4.97 _____ 0.497

28) 3.0504 _____ 3.0540

Convert Fraction to Decimal

Write each as a decimal.

1) $\frac{6}{10} =$

2) $\frac{58}{100} =$

3) $\frac{80}{100} =$

4) $\frac{5}{40} =$

5) $\frac{4}{50} =$

6) $\frac{7}{100} =$

7) $\frac{5}{80} =$

8) $\frac{21}{84} =$

9) $\frac{36}{400} =$

10) $\frac{3}{11} =$

11) $\frac{24}{48} =$

12) $\frac{18}{48} =$

13) $\frac{6}{20} =$

14) $\frac{9}{125} =$

15) $\frac{36}{120} =$

16) $\frac{15}{20} =$

17) $\frac{73}{100} =$

18) $\frac{9}{45} =$

19) $\frac{9}{10} =$

20) $\frac{6}{60} =$

21) $\frac{7}{42} =$

22) $\frac{11}{88} =$

Answer key Chapter 7

Graph Decimals

1) a. 0.432   b. 0.44   c. 0.444   d. 0.452

2) a. 0.577   b. 0.581   c. 0.585   d. 0592

3) a. 0.714   b. 0.718   c. 0.722   d. 0.731

4) a. 0.693   b. 0.70   c. 0.704   d. 0.713

Round Decimals

1) 0.9	11) 3.3	21) 5.41
2) 4.2	12) 4.2	22) 86.3
3) 8.8	13) 6.2	23) 354.3
4) 0.6	14) 9.16	24) 0.93
5) 8.0	15) 36.0	25) 0.007
6) 0.05	16) 48.0	26) 8.04
7) 9.4	17) 630.0	27) 41.66
8) 25.33	18) 577.8	28) 19.08
9) 4.63	19) 24.58	
10) 10.5	20) 84.0	

Decimals Addition

1) 63.35	6) 10.817	11) 50.481
2) 0.85	7) 97	12) 41.48
3) 33.99	8) 86.6	13) 51.82
4) 64.41	9) 54.2	14) 130.12
5) 31.11	10) 24.33	

Decimals Subtraction

1) 5.73	6) 20.73	11) 19.09
2) 30.07	7) 62.88	12) 11.01
3) 0.39	8) 26.08	13) 26.082
4) 16.554	9) 17.71	14) 136.64
5) 0.729	10) 8.099	

Decimals Multiplication

1) 4.56	2) 18.72	3) 8.2008

4) 459

5) 4,980

6) 174.552

7) 250.002

8) 13.8915

9) 1.2645

10) 206.1612

11) 180.544

12) 0.3824

13) 0.825

14) 92.862

Decimal Division

1) 0.007

2) 0.0003

3) 0.28

4) 0.0008

5) 0.2

6) 0.833….

7) 0.2

8) 0.05

9) 0.0009

10) 1.111…

11) 100

12) 2

13) 1.25

14) 1,000

15) 0.049

16) 0.073

17) 0.85

18) 3.68181…

19) 11.1818…

20) 10

21) 1.19428…

22) 0.3

23) 0.125

24) 0.00167

25) 10.467

26) 0.002

Comparing Decimals

1) <

2) >

3) =

4) <

5) >

6) <

7) >

8) <

9) >

10) >

11) <

12) <

13) <

14) >

15) >

16) =

17) <

18) <

19) <

20) >

21) <

22) >

23) >

24) >

25) >

26) =

27) >

28) <

Convert Fraction to Decimal

1) 0.6

2) 0.58

3) 0.8

4) 0.125

5) 0.08

6) 0.07

7) 0.0625

8) 0.25

9) 0.09

10) 0.27

11) 0.5

12) 0.375

13) 0.3

14) 0.072

15) 0.3

16) 0.75

17) 0.73

18) 0.2

19) 0.9

20) 0.1

21) 0.167

22) 0.125

Chapter 8:

Measurement

Reference Measurement

LENGTH

Customary	Metric
1 mile (mi) = 1,760 yards (yd)	1 kilometer (km) = 1,000 meters (m)
1 yard (yd) = 3 feet (ft)	1 meter (m) = 100 centimeters (cm)
1 foot (ft) = 12 inches (in.)	1 centimeter(cm) = 10 millimeters(mm)

VOLUME AND CAPACITY

Customary	Metric
1 gallon (gal) = 4 quarts (qt)	1 liter (L) = 1,000 milliliters (mL)
1 quart (qt) = 2 pints (pt.)	
1 pint (pt.) = 2 cups (c)	
1 cup (c) = 8 fluid ounces (Fl oz)	

WEIGHT AND MASS

Customary	Metric
1 ton (T) = 2,000 pounds (lb.)	1 kilogram (kg) = 1,000 grams (g)
1 pound (lb.) = 16 ounces (oz)	1 gram (g) = 1,000 milligrams (mg)

Time

1 year = 12 months
1 year = 52 weeks
1 week = 7 days
1 day = 24 hours
1 hour = 60 minutes
1 minute = 60 seconds

Metric Length Measurement

Convert to the units.

1) 2,000 mm = _____ cm

2) 5 m = _____ mm

3) 7 m = _____ cm

4) 9 km = _____ m

5) 5,000 mm = _____ m

6) 2,800 cm = _____ m

7) 13 m = _____ cm

8) 4,000 mm = _____ cm

9) 20,000 mm = _____ m

10) 7 km = _____ mm

11) 6 km = _____ m

12) 3 m = _____ cm

13) 17,000 m = _____ km

14) 500,000 m = _____ km

Customary Length Measurement

Convert to the units.

1) 15 ft = _____ in

2) 8 ft = _____ in

3) 7 yd = _____ ft

4) 9 yd = _____ ft

5) 3 yd = _____ in

6) 3 mi = _____ in

7) 7,200 in = _____ yd

8) 252 in = _____ yd

9) 8,800 yd = _____ mi

10) 12 yd = _____ in

11) 4 mi = _____ yd

12) 47,520 ft = _____ mi

13) 60 in = _____ ft

14) 25 yd = _____ ft

15) 36 in = _____ ft

16) 2 mi = _____ ft

Metric Capacity Measurement

Convert the following measurements.

1) 50 l = _____ ml

2) 4 l = _____ ml

3) 13 l = _____ ml

4) 8 l = _____ ml

5) 19 l = _____ ml

6) 2 l = _____ ml

7) 70,000 ml = _____ l

8) 8,000 ml = _____ l

9) 37,000 ml = _____ l

10) 200,000 ml = _____ l

11) 6,000,000 ml = _____ l

12) 40,000 ml = _____ l

Customary Capacity Measurement

Convert the following measurements.

1) 2 gal = _____ qt.

2) 11 gal = _____ pt.

3) 3 gal = _____ c.

4) 14 pt. = _____ c

5) 43 c = _____ fl oz

6) 16 qt = _____ pt.

7) 8 qt = _____ c

8) 29 pt. = _____ c

9) 6,720 c = _____ gal

10) 144 pt. = _____ gal

11) 72 qt = _____ gal

12) 92 pt. = _____ qt

13) 4,600 c = _____ qt

14) 146 c = _____ pt.

15) 108 qt = _____ gal

16) 1,848 pt. = _____ qt

17) 31 gal = _____ pt.

18) 6 qt = _____ c

19) 640 c = _____ gal

20) 104 fl oz = _____ c

Metric Weight and Mass Measurement

Convert.

1) 7 kg = _____ g

2) 3 kg = _____ g

3) 13 kg = _____ g

4) 21 kg = _____ g

5) 9 kg = _____ g

6) 121 kg = _____ g

7) 249 kg = _____ g

8) 4,000 g = _____ kg

9) 6,000 g = _____ kg

10) 17,000 g = _____ kg

11) 129,000 g = _____ kg

12) 220,000 g = _____ kg

13) 9,000,000 g = _____ kg

14) 11,000,000 g = _____ kg

Customary Weight and Mass Measurement

Convert.

1) 16,000 lb. = _____ T

2) 20,000 lb. = _____ T

3) 170,000 lb. = _____ T

4) 44,000 lb. = _____ T

5) 7 lb. = _____ oz

6) 4 lb. = _____ oz

7) 10 lb. = _____ oz

8) 24 T = _____ lb.

9) 3 T = _____ lb.

10) 9 T = _____ lb.

11) 112 T = _____ lb.

12) 2 T = _____ oz

13) 5 T = _____ oz

14) 224 oz = _____ lb.

Time

Convert to the units.

1) 16 hr. = _____ min

2) 9 year = _____ week

3) 2 hr. = _____ sec

4) 12 min = _____ sec

5) 600 min = _____ hr

6) 730 day = _____ year

7) 3 year = _____ hr.

8) 27 day = _____ hr

9) 4 day = _____ min

10) 540 min = _____ hr

11) 16 year = _____ month

12) 19,200 sec = _____ min

13) 288 hr = _____ day

14) 23 weeks = _____ day

How much time has passed?

15) From 4:15 A.M. to 7:25 A.M.: ____ hours and ___ minutes.

16) From 3:40 A.M. to 8:25 A.M.: ____ hours and ___ minutes.

17) It's 8:50 P.M. What time was 2 hours ago? _____ O'clock

18) 3:20 A.M to 6:40 AM: _____ hours and _____ minutes.

19) 3:30 A.M to 6:05 AM: _____ hours and _____ minutes.

20) 7:10 A.M. to 8:15 AM. = _____ hour(s) and _____ minutes.

21) 11:55 A.M. to 4:25 PM. = _____ hour(s) and _____ minutes

22) 7:18 A.M. to 7:52 A.M. = _____ minutes

23) 9:13 A.M. to 9:50 A.M. = _____ minutes

Answers of Worksheets – Chapter 8

Metric length

1) 200 cm
2) 5,000 mm
3) 700 cm
4) 9,000 m
5) 5 m
6) 28 m
7) 1,300 cm
8) 400 cm
9) 20 m
10) 7,000,000 mm
11) 6,000 m
12) 300 cm
13) 17 km
14) 500 km

Customary Length

1) 180
2) 96
3) 21
4) 27
5) 108
6) 190,080
7) 200
8) 7
9) 5
10) 432
11) 7,040
12) 9
13) 5
14) 75
15) 3
16) 10,560

Metric Capacity

1) 50,000 ml
2) 4,000 ml
3) 13,000 ml
4) 8,000 ml
5) 19,000 ml
6) 2,000 ml
7) 70 L
8) 8 L
9) 37 L
10) 200 L
11) 6,000 L
12) 40 L

Customary Capacity

1) 8 qt
2) 88 pt.
3) 48 c
4) 28 c
5) 344 fl oz
6) 12 pt.
7) 32 c
8) 58 c
9) 420 gal
10) 18 gal
11) 18 gal
12) 46 qt
13) 1,150qt
14) 73 pt.
15) 27 gal
16) 927 qt
17) 248 pt.
18) 24 c
19) 40 gal
20) 13 c

Metric Weight and Mass

1) 7,000 g
2) 3,000 g
3) 13,000 g
4) 21,000 g
5) 9,000 g
6) 121,000 g
7) 249,000 g
8) 4 kg
9) 6 kg
10) 17 kg
11) 129 kg
12) 220 kg
13) 9,000 kg
14) 11,000 kg

Customary Weight and Mass

1) 8 T	6) 64 oz	11) 224,000 lb.
2) 10 T	7) 160 oz	12) 64,000 oz
3) 85 T	8) 48,000 lb.	13) 160,000 oz
4) 22 T	9) 6,000 lb.	14) 14 lb
5) 112 oz	10) 18,000 lb.	

Time

1) 960 min	9) 5,760 min	17) 6:50 P.M.
2) 468 weeks	10) 9 hr	18) 3:20
3) 7,200 sec	11) 192 months	19) 2:35
4) 720 sec	12) 320 min	20) 1:05
5) 10 hr	13) 12 days	21) 4:30
6) 2 year	14) 161 days	22) 34 minutes
7) 26,280 hr	15) 3:10	23) 37 minutes
8) 648 hr	16) 4:45	

Chapter 9: Symmetry and Transformations

Line Segments

Write each as a line, ray, or line segment.

1)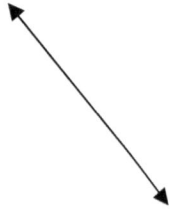

2)

3)

4)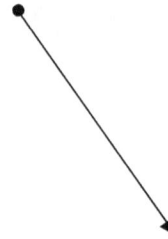

5)

6)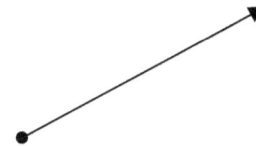

7)

8)

Parallel, Perpendicular and Intersecting Lines

State whether the given pair of lines are parallel, perpendicular, or intersecting.

1)

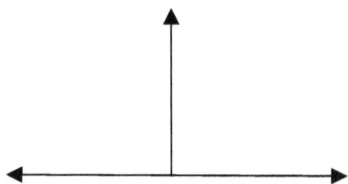

2)

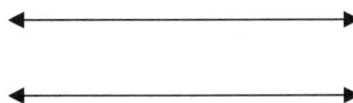

3)

4)

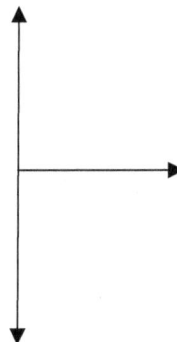

5)

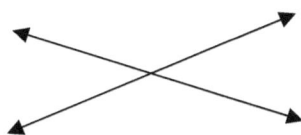

6)

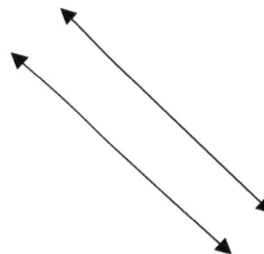

7)

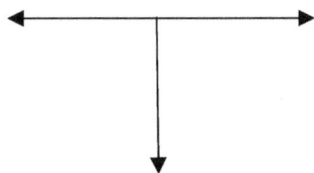

8)

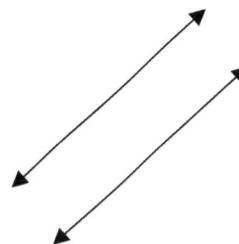

Identify Lines of Symmetry

Tell whether the line on each shape a line of symmetry is.

1)

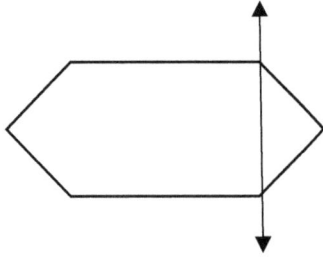

2)

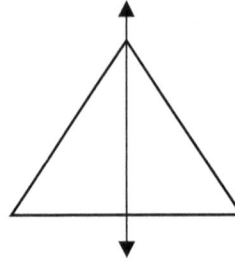

3)

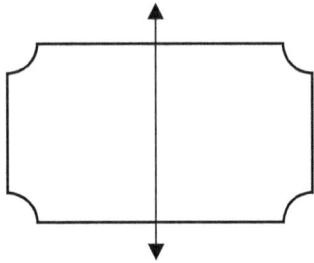

4)

5)

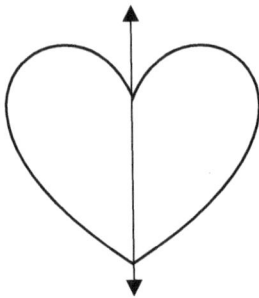

6)

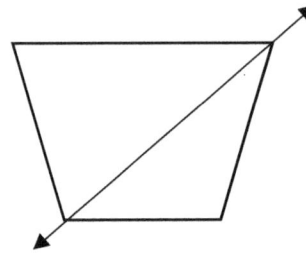

7)

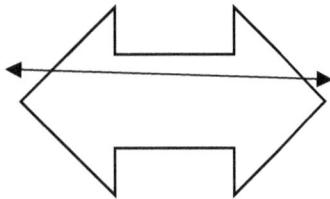

8)

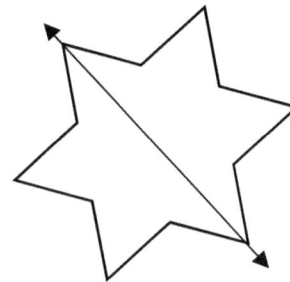

Lines of Symmetry

Draw lines of symmetry on each shape. Count and write the lines of symmetry you see.

1)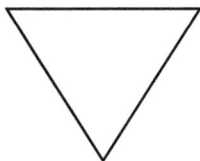

2)

3)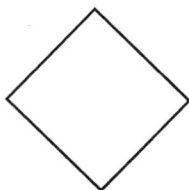

4)

5)

6)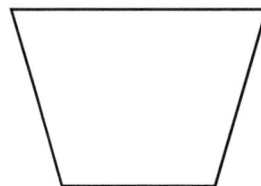

7)

8)

Identify Three–Dimensional Figures

Write the name of each shape.

1)

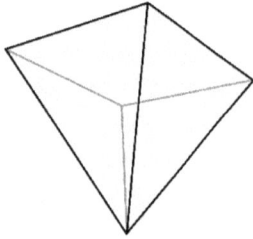

2)

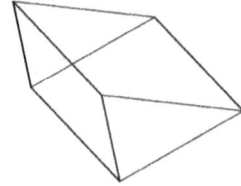

3)

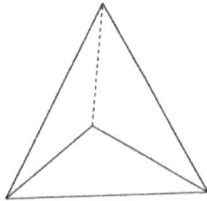

4)

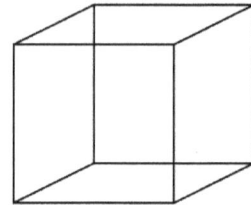

5)

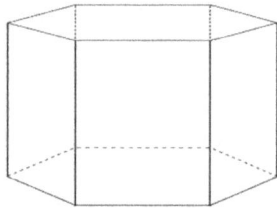

6)

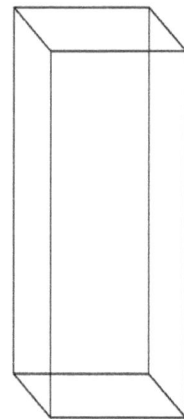

7)

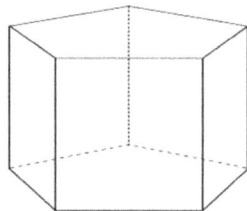

Vertices, Edges, and Faces

Complete the chart below.

Shape	Number of edges	Number of faces	Number of vertices
1)	_____	_____	_____
2)	_____	_____	_____
3)	_____	_____	_____
4)	_____	_____	_____
5)	_____	_____	_____
6)	_____	_____	_____

Identify Faces of Three–Dimensional Figures

Write the number of faces.

1)

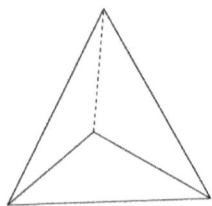

2)

3)

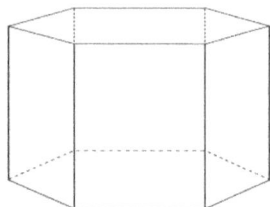

4)

5)

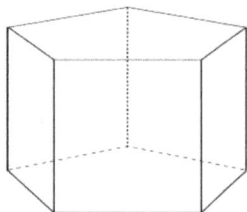

6)

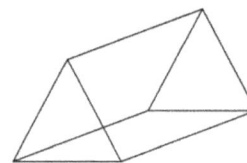

7)

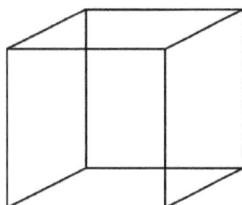

8)

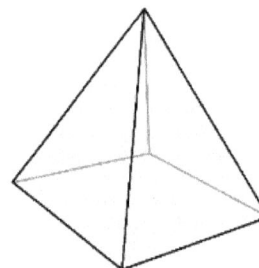

Answers of Worksheets – Chapter 9

Line Segments

1) Line

2) Line segment

3) Line segment

4) Ray

5) Line

6) Ray

7) Line segment

8) Ray

Parallel, Perpendicular and Intersecting Lines

1) Perpendicular

2) Parallel

3) Intersection

4) Perpendicular

5) Intersection

6) Parallel

7) Perpendicular

8) Parallel

Identify lines of symmetry

1) No

2) yes

3) yes

4) No

5) yes

6) No

7) No

8) yes

lines of symmetry

1)

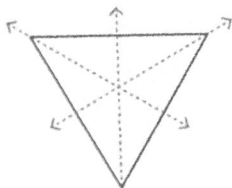

2)

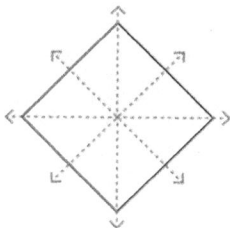

3)

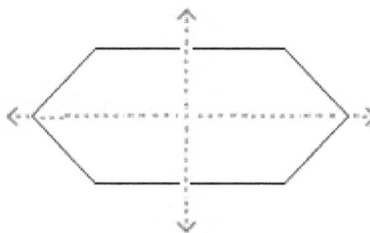

4)

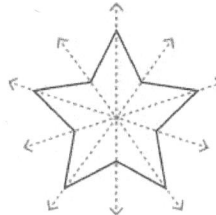

5)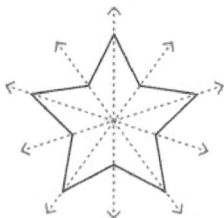

6)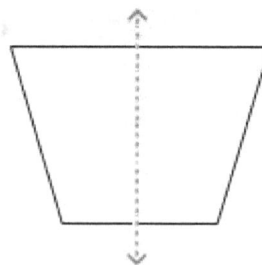

7)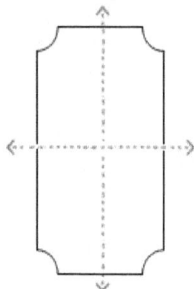

8)

Identify Three–Dimensional Figures

1) Square pyramid

2) Triangular prism

3) Triangular pyramid

4) Cube

5) Hexagonal prism

6) Rectangular prism

7) Pentagonal prism

Vertices, Edges, and Faces

	Shape	Number of edges	Number of faces	Number of vertices
1)		18	8	12
2)		12	6	8
3)		6	4	4

4) 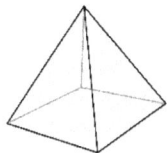 8   5   5

5) 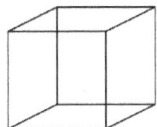 12   6   8

6) 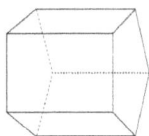 15   7   10

Identify Faces of Three–Dimensional Figures

1) 4   4) 2   7) 6
2) 6   5) 7   8) 5
3) 8   6) 5

Chapter 10:

Geometry

Identifying Angles

Write the name of the angles(Acute, Right, Obtuse, and Straight Angles) .

1)

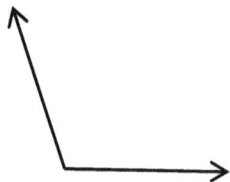

2)

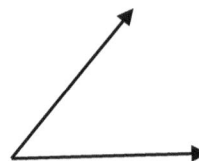

3)

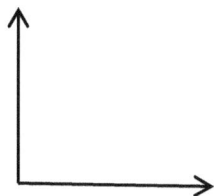

4)

5)

6)

7)

8)

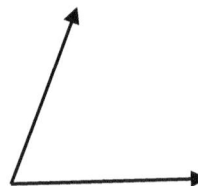

Polygon Names

Write name of polygons.

1)

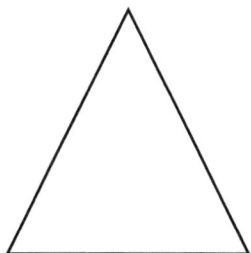

2)

3)

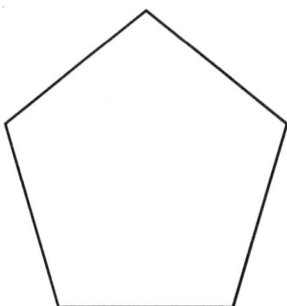

4)

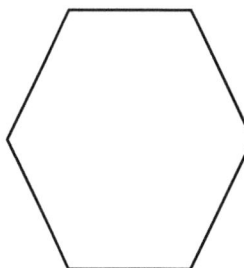

5)

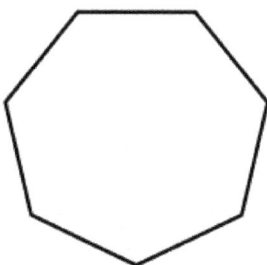

6)

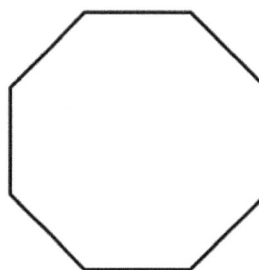

7)

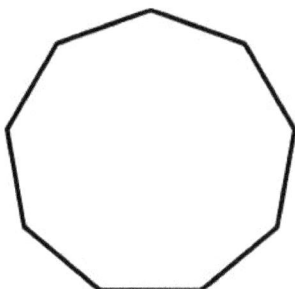

8)

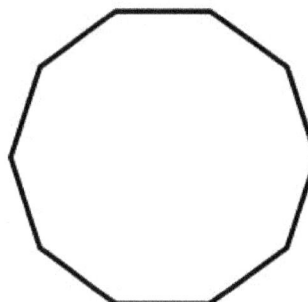

Triangles

Classify the triangles by their sides and angles.

1)

2)

3)

4)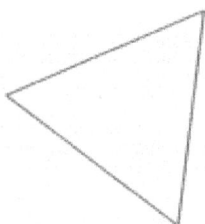

5)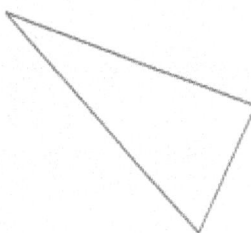

6)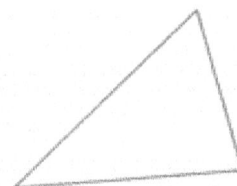

Find the measure of the unknown angle in each triangle.

7)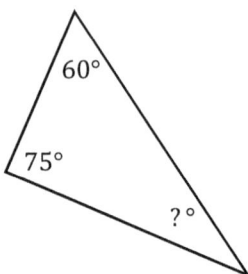

8)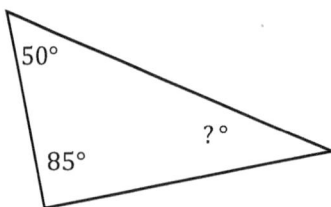

9)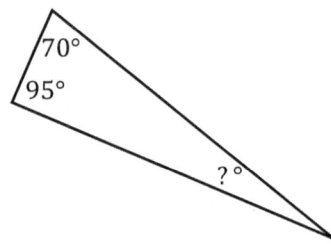

10)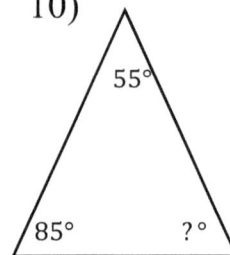

11)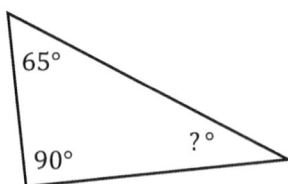

12)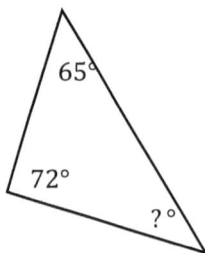

13)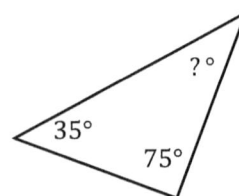

14)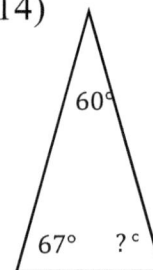

Quadrilaterals and Rectangles

Write the name of quadrilaterals.

1)

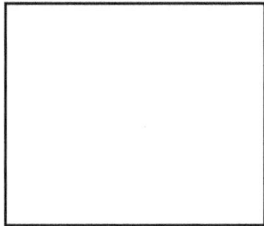

2)

3)

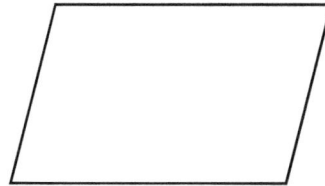

4)

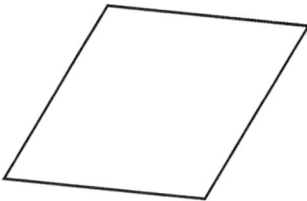

5)

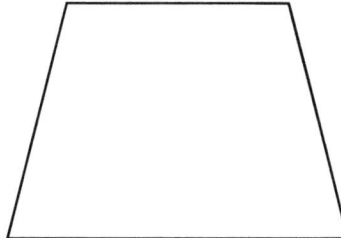

6)

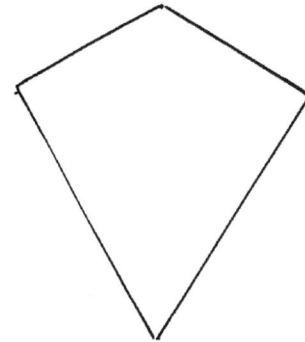

Solve.

7) A rectangle has _____ sides and _____ angles.

8) Draw a rectangle that is 6 centimeters long and 5 centimeters wide. What is the perimeter?

9) Draw a rectangle 5 cm long and 3 cm wide.

10) Draw a rectangle whose length is 6cm and whose width is 4 cm. What is the perimeter of the rectangle?

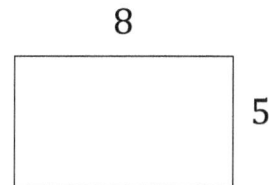

11) What is the perimeter of the rectangle?

Area and Perimeter of Square

Find the perimeter and area of each squares.

1)

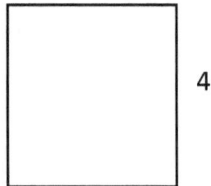

Perimeter:............:

Area:..........:

2)

Perimeter:...............:

Area:...............:

3)

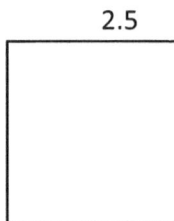

Perimeter:...............:

Area:.............:

4)

Perimeter:...............:

Area:..............:

5)

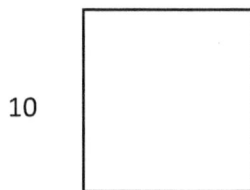

Perimeter:............:

Area:............:

6)

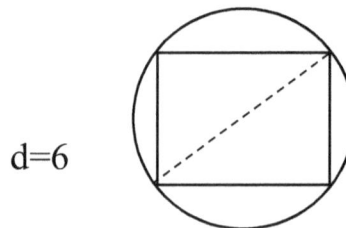

Perimeter of Square:..............:

Area of Square:...........:

Area and Perimeter of Rectangle

Find the perimeter and area of each rectangle.

1)

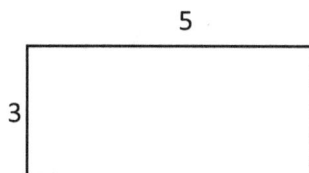

Perimeter:_____:

Area:_____:

2)

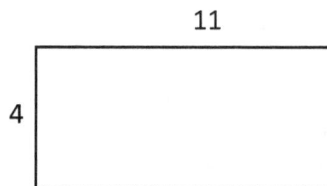

Perimeter:_____.

Area:_____:

3)

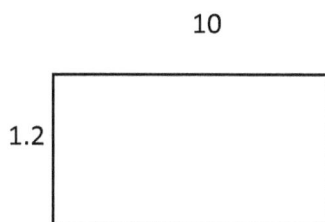

Perimeter:_____:

Area:_____:

4)

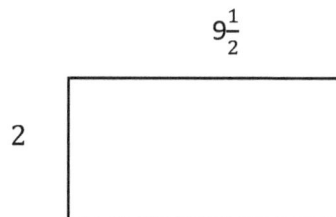

Perimeter:_____:

Area:_____:

5)

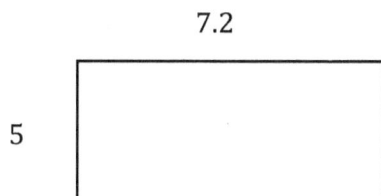

Perimeter:_____.

Area:_____:

6)

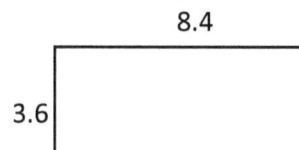

Perimeter:_____.

Area:_____:

Area and Perimeter of Triangle

Find the perimeter and area of each triangle.

1)

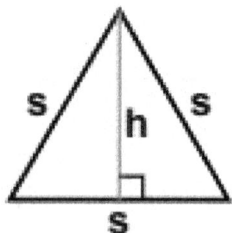

Perimeter:_____.

Area:_____.

2)

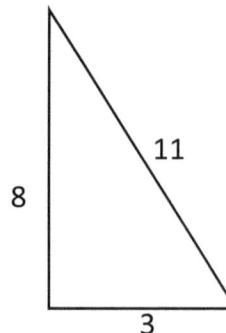

11

8

3

Perimeter:_____:

Area:_____:

3)

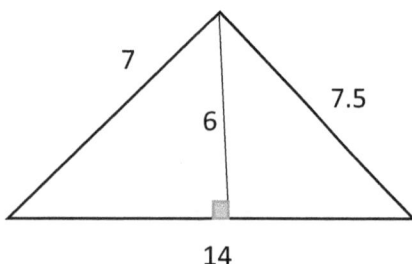

7

6

7.5

14

Perimeter:_____:

Area: _____:

4)

s=10

h=6.4

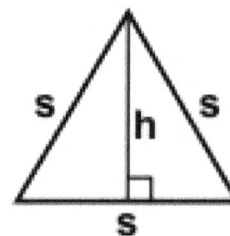

Perimeter:_____:

Area:_____:

5)

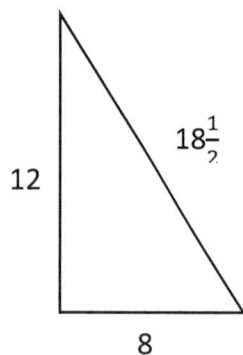

$18\frac{1}{2}$

12

8

Perimeter:_____:

Area:_____:

6)

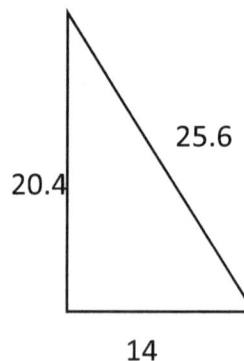

25.6

20.4

14

Perimeter:_____:

Area:_____:

Perimeter of Polygon

Find the perimeter of each polygon.

1)

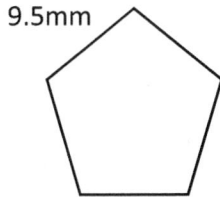

9.5mm

Perimeter:_____.

2)

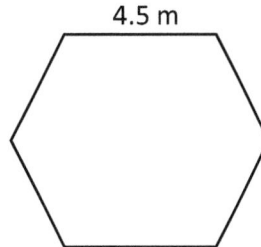

4.5 m

Perimeter:_____:

3)

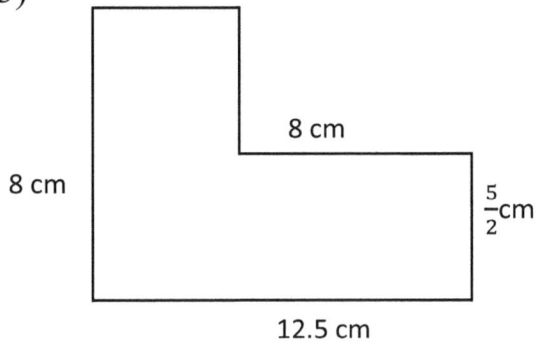

8 cm

8 cm

$\frac{5}{2}$cm

12.5 cm

Perimeter:_____.

4)

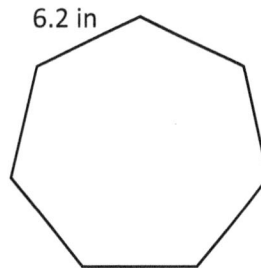

6.2 in

Perimeter:_____:

5)

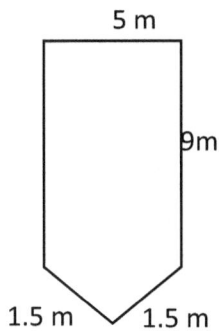

5 m

9m

1.5 m   1.5 m

Perimeter:_____.

6)

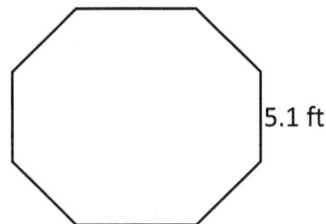

5.1 ft

Perimeter:_____:

Answer key Chapter 10

Identifying Angles

1) Obtuse
2) Acute
3) Right
4) Acute
5) Straight
6) Obtuse
7) Obtuse
8) Acute

Polygon Names

1) Triangle
2) Quadrilateral
3) Pentagon
4) Hexagon
5) Heptagon
6) Octagon
7) Nonagon
8) Decagon

Triangles

1) Scalene, obtuse
2) Isosceles, right
3) Scalene, right
4) Equilateral, acute
5) Scalene, acute
6) Scalene, acute
7) $45°$
8) $45°$
9) $15°$
10) $40°$
11) $25°$
12) $43°$
13) $70°$
14) $53°$

Quadrilaterals and Rectangles

1) Square
2) Rectangle
3) Parallelogram
4) Rhombus
5) Trapezoid
6) Kike
7) 4 - 4
8) 22
9) Use a rule to draw the rectangle
10) 20
11) 26

Area and Perimeter of Square

1. Perimeter: 16, Area:16
2. Perimeter: 6, Area:2.25
3. Perimeter: 10, Area:6.25
4. Perimeter: 32, Area:64
5. Perimeter: 40, Area:100
6. Perimeter: $4\sqrt{3}$, Area:3

Area and Perimeter of Rectangle

1- Perimeter: 16, Area:15
2- Perimeter: 30, Area:44
3- Perimeter: 22.4, Area:12
4- Perimeter: 23, Area:19
5- Perimeter: 24.4, Area:36
6- Perimeter:24, Area:30.24

Area and Perimeter of Triangle

1- Perimeter: 3s, Area:$\frac{1}{2}sh$
2- Perimeter: 22, Area:12
3- Perimeter: 28.5, Area:42
4- Perimeter: 30, Area:32
5- Perimeter: 38.5, Area:48
6- Perimeter: 60, Area:142.8

Perimeter of Polygon

1) 47.5 mm
2) 27 m
3) 41 cm
4) 43.4 in
5) 26 m
6) 40.8 ft

Chapter 11: Data and Graphs

Tally and Pictographs

Using the key, draw the pictograph to show the information.

🖥️	||||
🖱️	卌 卌 ||
💾	卌 |
🎧	|||
🔌	卌 |||

🖥️	
🖱️	
💾	
🎧	
🔌	

Key: ⁑ = 2 Hardware

Stem–And–Leaf Plot

Make stem-and-leaf plots for the given data.

1) 15, 16, 38, 31, 12, 54, 18, 37, 39, 34, 19, 32, 55

Stem	leaf

2) 72, 74, 17, 41, 72, 14, 46, 78, 48, 44, 49, 42

Stem	leaf

3) 125, 108, 65, 65, 105, 127, 62, 126, 68, 124, 66, 109

Stem	leaf

4) 61, 45, 66, 60, 99, 63, 90, 97, 68, 63, 49, 42

Stem	leaf

5) 55, 58, 105, 56, 15, 108, 102

Stem	leaf

6) 123, 57, 77, 55, 120, 127, 73, 124, 58, 123, 79, 71

Stem	leaf

Dot plots

The ages of students in a Math class are given below.

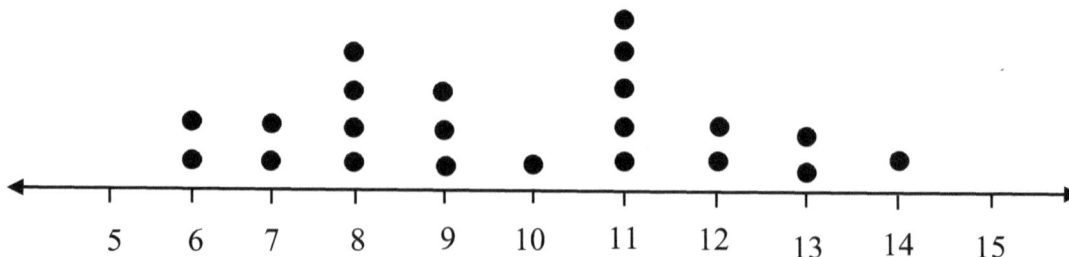

1) What is the total number of students in math class?

2) How many students are at least 12 years old?

3) Which age(s) has the most students?

4) Which age(s) has the fewest student?

5) Determine the median of the data.

6) Determine the range of the data.

7) Determine the mode of the data.

Coordinate Plane

Plot each point on the coordinate grid.

1) A $(2, 7)$

2) B $(6, 3)$

3) C $(0, 7)$

4) D $(3, 0)$

5) E $(1, 4)$

6) F $(3, 9)$

7) G $(5, 1)$

8) H $(7, 7)$

9) I $(9, 6)$

10) J $(6, 1)$

11) K $(2, 4)$

12) L $(3, 8)$

Bar Graph

Each student in class selected two games that they would like to play. Graph the given information as a bar graph and answer the questions below:

Game	Votes
Football	13
Volleyball	10
Basketball	18
Baseball	17
Tennis	13

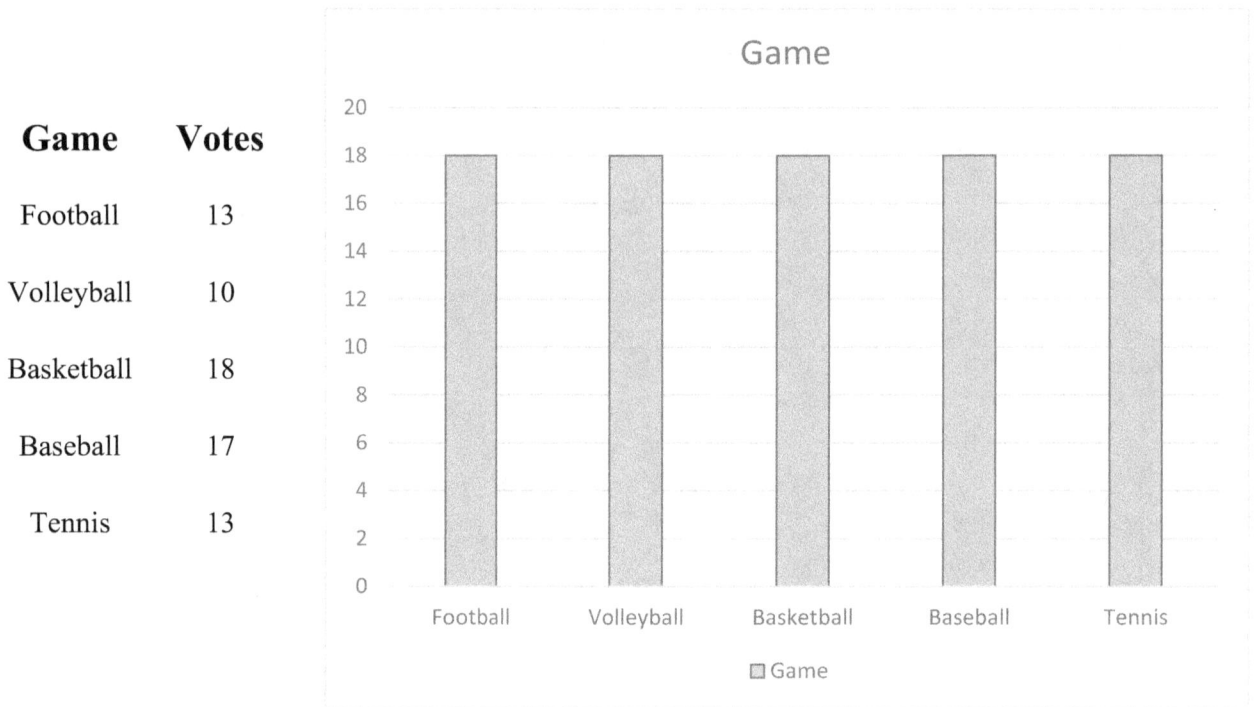

1) Which was the most popular game to play?

2) How many more students like Basketball than Volleyball?

3) Which two game got the same number of votes?

4) How many Volleyball and Football did student vote in all?

5) Did more student like football or Volleyball?

6) Which game did the fewest student like?

Line Graphs

Amelia works in a doll store. She records the number of dolls sold in five days on a line graph. Use the graph to answer the questions.

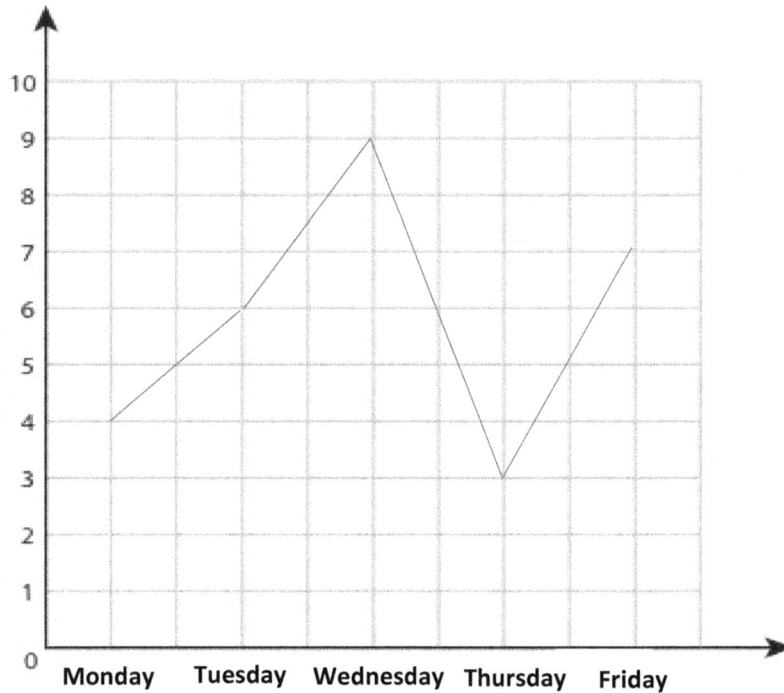

1) How many dolls were sold on Tuesday?

2) Which day had the minimum sales of dolls?

3) Which day had the maximum number of dolls sold?

4) How many dolls were sold in 5 days?

Answer key Chapter 11

Tally and Pictographs

🖥	✳ ✳
🖱	✳ ✳ ✳ ✳ ✳ ✳
💾	✳ ✳ ✳
🎧	✳ ◁
🔌	✳ ✳ ✳ ✳

Stem–And–Leaf Plot

1)

Stem	leaf
1	2 5 6 8 9
3	1 2 4 7 8 9
5	4 5

2)

Stem	leaf
1	4 7
4	1 2 4 6 8 9
7	2 2 4 8

3)

Stem	leaf
6	2 5 5 6 8
10	5 8 9
12	4 5 6 7

4)

Stem	leaf
4	2 9 5
6	0 1 3 3 6 8
9	0 7 9

5)

Stem	leaf
1	5
5	5 6 8
10	2 5 8

6)

Stem	leaf
5	5 7 8
7	1 3 7 9
12	0 3 3 4 7

Dot plots

1) 22

2) 5

3) 11

4) 10 and 14

5) 2

6) 4

7) 2

Graph Points on a Coordinate Plane

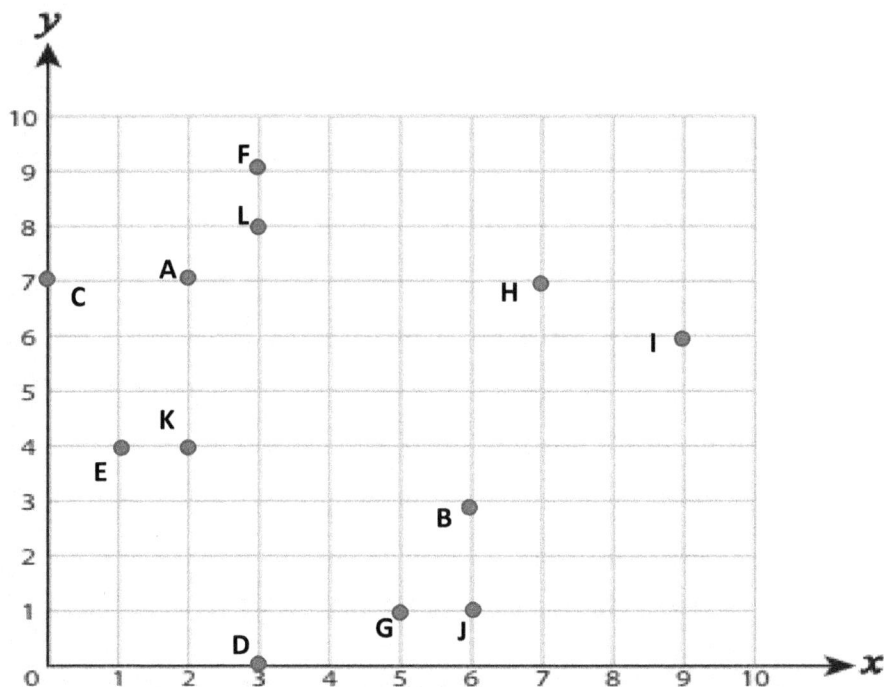

Bar Graph

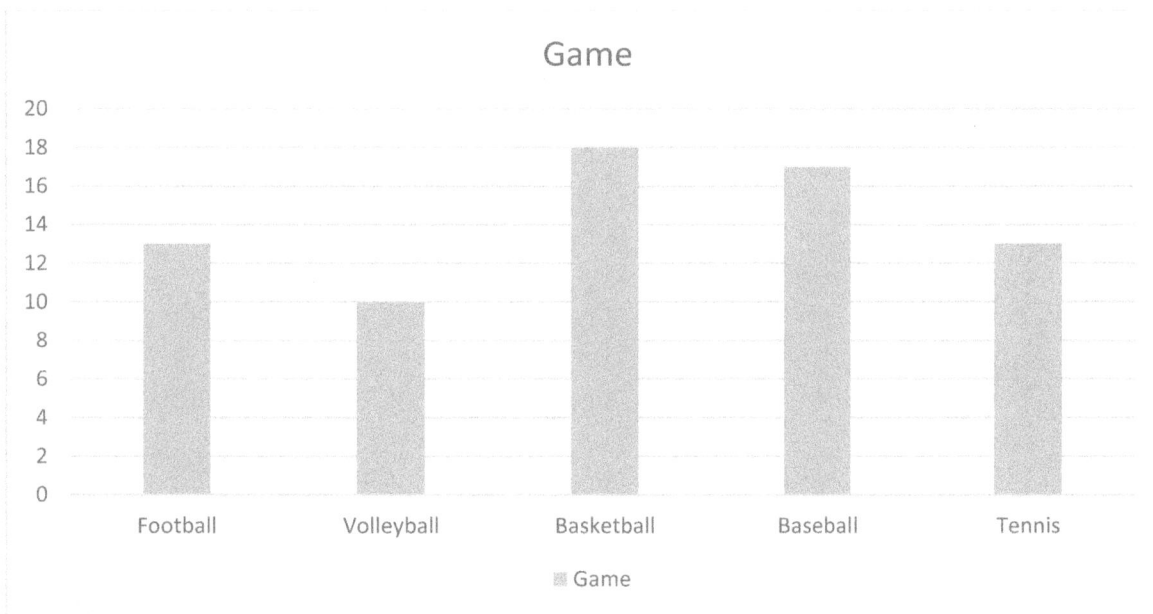

1) Basketball

2) 8 students

3) Football and Tennis

4) 23

5) Football

6) Volleyball

Line Graphs

1) 6

2) Thursday

3) Wednesday

4) 29

FSA

Test Review

FSA GRADE 4 MAHEMATICS REFRENCE MATERIALS

LENGTH

Customary	Metric
1 mile (mi) = 1,760 yards (yd)	1 kilometer (km) = 1,000 meters (m)
1 yard (yd) = 3 feet (ft)	1 meter (m) = 100 centimeters (cm)
1 foot (ft) = 12 inches (in.)	1 centimeter (cm) = 10 millimeters (mm)

VOLUME AND CAPACITY

Customary	Metric
1 gallon (gal) = 4 quarts (qt)	1 liter (L) = 1,000 milliliters (mL)
1 quart (qt) = 2 pints (pt.)	
1 pint (pt.) = 2 cups (c)	
1 cup (c) = 8 fluid ounces (Fl oz)	

WEIGHT AND MASS

Customary	Metric
1 ton (T) = 2,000 pounds (lb.)	1 kilogram (kg) = 1,000 grams (g)
1 pound (lb.) = 16 ounces (oz)	1 gram (g) = 1,000 milligrams (mg)

Time

1 year = 12 months	1 day = 24 hours
1 year = 52 weeks	1 hour = 60 minutes
1 week = 7 days	1 minute = 60 seconds

Perimeter

Square	$P = 4S$
Rectangle	$P = L + W + L + W$ or $P = 2L + 2W$

Area

Square	$A = S \times S$
Rectangle	$A = L \times W$

The Florida Standards
Assessments
FSA Practice Test 1

Mathematics

GRADE 4

Administered *Month Year*

Session 1

❖ Calculators are NOT permitted for this practice test.

❖ Time for the Session: 80 Minutes

1) Which decimal is equivalent to $\frac{26}{100}$?

    A. 0.026

    B. 0.26

    C. 2.6

    D. 26.00

2) In which drawing does line A appear to be perpendicular to line B?

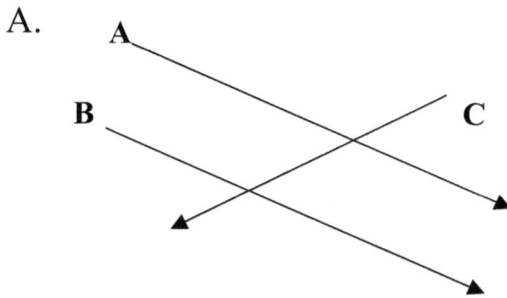

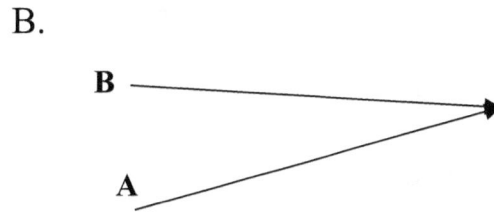

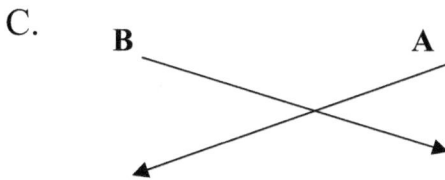

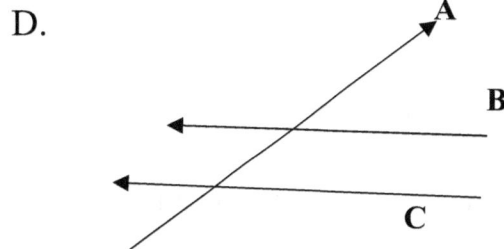

3) Which digit can replace the box below to make the comparing true?

$$7,632 < 7,\boxed{}19 < 7,801$$

    A. 4

    B. 7

    C. 8

    D. 9

4) Patricia is having a party. A total of 43 people is coming. How many pizzas will she need to buy if each person has 2 slices, and the pizzas are sliced into sixths?

A. 15

B. 22

C. 17

D. 86

5) What is the value of expression $2,524 \times 5$?

A. 10,620

B. 10,560

C. 15,260

D. 12,620

6) Emily writes a number.

- The number is between 60 and 70.

- It has exactly 3 factors.

- One of the factors is 11.

- What is Emily's number

A. 66

B. 68

C. 62

D. 72

7) Nancy makes mango smoothies for a party. She combines 5 gallons 2 quarts of milk with 3 gallons 3 quarts of mango juice. Her sister adds 2 gallons 2 quarts of the smoothies. How much is all the mango smoothies?

A. 10 gallons

B. 11 gallons 2 quarts

C. 11 gallons 3 quarts

D. 10 gallons 2 quarts

8) Emma had 8 feet of yarn. She needed 6 inches to make a bracelet. How many bracelets can she make with the amount of yarn she has?

A. 16

B. 36

C. 38

D. 6

9) Sam has 3,600 dimes. Anna has $\frac{1}{20}$ number of dimes that Sam does. George has $\frac{1}{5}$ the number of dimes that Anna has. How much money does George have?

A. $0.36

B. $3.60

C. $36

D. $240

10) Which expression can be used to solve the equation $6,300 \div 9 = $ ___?___ .

    A. $(63 \div 9) + (600 \div 9)$

    B. $(600 \div 9) - (300 \div 9)$

    C. $(6,000 \div 9) + (300 \div 9)$

    D. $(6,000 \div 9) - (300 \div 9)$

11) A company used 53,049.001 feet of electrical wires last month. What is the expanded form of the number?

    A. $(5 \times 10,000) + (3 \times 1,000) + (4 \times 100) + (9 \times 10) + (1 \times 0.01)$

    B. $(5 \times 10,000) + (3 \times 1,000) + (4 \times 10) + (9 \times 1) + (1 \times 0.01)$

    C. $(5 \times 10,000) + (3 \times 100) + (4 \times 10) + (9 \times 1) + (1 \times 0.001)$

    D. $(5 \times 10,000) + (3 \times 1,000) + (4 \times 10) + (9 \times 1) + (1 \times 0.001)$

12) Select these polygons that have at least one set of parallel sides.

    Figure 1   Figure 2   Figure 3   Figure 4

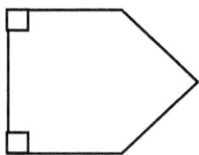  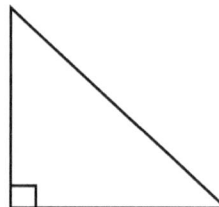 

    A. Figure 1 and Figure 3

    B. Figure 3 and Figure 4

    C. Figure 1, Figure 2, and Figure 4

    D. All figures.

13) The width of a rectangular yard is 55 feet, and the length is 120 feet. A tent that is 22 feet long and 18 feet wide cover part of the field. How many square feet of the field are not covered by the tent?

A. 6,420 square feet

B. 6,204 square feet

C. 4,620 square feet

D. 396 square feet

14) In which model could the shaded parts equivalent to $1 \times \frac{5}{6}$?

A.

B.

C.

D.

15) Russel has 6 acres of land and he grows vegetables on one-fifth of it. Philip has 13 acres and one-sixth of it is planted with vegetables. Who has the biggest area of vegetables?

A. Need more information.

B. Russel

C. Philip

D. They have the same area.

Session 2

❖ **Calculators are NOT permitted for this practice test.**

❖ **Time for the Session: 80 Minutes**

16) Which two numbers round to 6,700 when rounded to the nearest hundreds?

    A. 6,471 and 6,667

    B. 6,532 and 6,638

    C. 6,681 and 6,721

    D. 6,576 and 6,648

17) Which expression has a quotient of about 7?

    A. $10 \div 3$

    B. $39 \div 8$

    C. $71 \div 9$

    D. $43 \div 7$

18) Which number is greater than 24.45 and less than 24.71?

    A. 24.96

    B. 24.44

    C. 24.73

    D. 24.53

19) Which figure represented a ray?

A.

B.

C.

D.

20) The Ethan's family are on the road trip. They travel 86.24 miles the first day, 96.53 miles the second day, and 112.38 miles the final day. How many miles does the Ethan's family travel during the three-day trip?

A. 308.91 miles

B. 208.91 miles

C. 295.15 miles

D. 276.15 miles

21) Maria starts practicing her guitar at 11:23 A.M. she practices for 1 hours and 45 minutes and then stop 53 minutes to eat lunch. What time does Maria lunch end?

   i.   13:16 P.M.

  ii.   13:58 P.M.

 iii.   02:10 A.M.

 iv.   14:01 P.M.

22) What is the measure of angle DOC to the nearest degree?

A. 70°

B. 60°

C. 110°

D. 130°

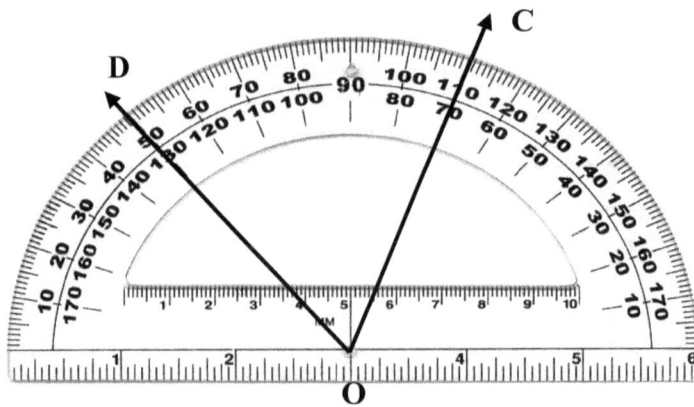

23) The data shows the 30 hours rain by each of 5 days. Which frequency table represents the number of hours rain each day?

3 hours rain Monday	4 fewer hours rain Wednesday than Tuesday.
8 hours rain Tuesday	3 times as many as Monday rain Friday.

The rest of hours rain Thursday.

B.

A.

Hours of Rain

Day	Number of hours
Monday	|||
Tuesday	||||| |||
Wednesday	||||| |||
Thursday	|||||
Friday	|||||	

B.

Hours of Rain

Day	Number of hours
Monday	|||
Tuesday	|||||	
Wednesday	||||| |||
Thursday	
Friday	||||| ||||

C.

Hours of Rain

Day	Number of hours
Monday	|||
Tuesday	||||| |||
Wednesday	||||
Thursday	|||||	
Friday	||||| ||||

D.

Hours of Rain

Day	Number of hours
Monday	|||
Tuesday	||||| ||
Wednesday	||||
Thursday	|||||	
Friday	||||| ||

24) A baker sold 11 cakes for the same price. He received a total of $440. Each cake

cost the baker $25 to make. How much money did he make on each cake?

A. $40

B. $15

C. $25

D. $33

25) An equation that uses fraction model is shown. Which fraction makes the

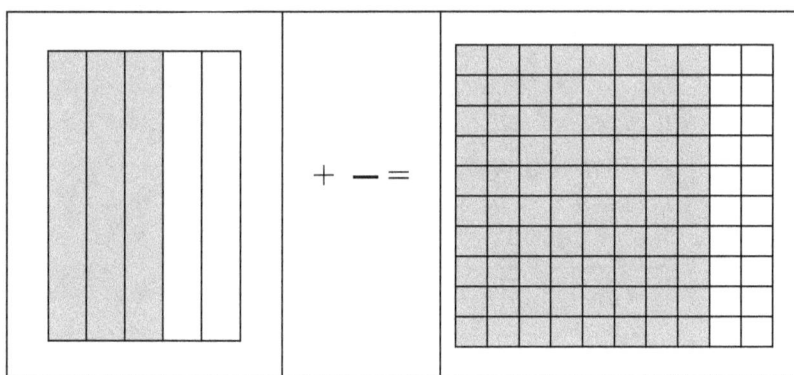

equation true?

A. $\frac{3}{80}$

B. $\frac{2}{5}$

C. $\frac{1}{5}$

D. $\frac{3}{10}$

$+ - =$

26) Ava is making a pattern for quilt. The pattern shows 60 squares. Every 6th square

is green. How many green squares are in the pattern?

A. 6

B. 12

C. 18

D. 30

27) There are 315 math students in the 5th grade. There are 7 math classes of equal

size. How many math students are in each class?

A. 120

B. 30

C. 45

D. 63

28) Emily collected 33 coins; Ella collected 52 coins. Zoey collected 44 coins. If

they share all the coins equally, how many coins will each person get?

A. 96

B. 43

C. 45

D. 130

29) Natalie has $4\frac{3}{4}$ gallon of orange juice. She uses $2\frac{1}{4}$ gallon for a party. How many

gallons of orange juice does Natalie have left?

A. $2\frac{3}{4}$

B. $1\frac{1}{4}$

C. 6

D. $2\frac{1}{2}$

30) How many lines of symmetry does the shape shown have?

A. 2

B. 1

C. 4

D. 0

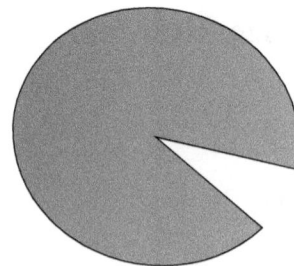

The Florida Standards Assessments

FSA Practice Test 2

Mathematics

GRADE 4

Administered *Month Year*

Session 1

❖ **Calculators are NOT permitted for this practice test.**

❖ **Time for the Session: 80 Minutes**

1) Grace writes a number.

- The digits in ones period are 304

- The word form of the thousands period is nine hundred six thousand.

- The digit in tens millions and tens place are the same.

What is the Grace's number?

A. 304,960,039

B. 304,906,309

C. 906,309,304

D. 309,906,304

2) What is the quotient for the expression $3,284 \div 8$?

A. 401

B. 410 r 8

C. 410 r 4

D. 401 r 8

3) Which fraction equivalent to 8.2?

A. $\frac{82}{10}$

B. $\frac{82}{100}$

C. $\frac{10}{82}$

D. $\frac{100}{82}$

4) Thomas is adding a baseboard trim around the edge of his floor. The room is 18 feet wide and 30 feet long. There is a door opening that is 4 feet wide. How many feet of trim will Thomas need?

A. 85 feet

B. 92 feet

C. 540 feet

D. 524 feet

5) Which of the following fractions is higher than $\frac{1}{12}$, but less than $\frac{2}{5}$?

A. $\frac{3}{10}$

B. $\frac{1}{20}$

C. $\frac{1}{10}$

D. $\frac{2}{3}$

6) Taye had four $7 bill, five nickels, and 12 pennies. Then he bout a math book for $20.68. How much money did Taye have after bout the book?

A. $7.32

B. $7.57

C. $7.69

D. $7.96

7) William has a rope 8 feet 12 inches long. He cuts it into 12 equal pieces. How many inches long is each piece?

A. 9

B. 6

C. 11

D. 10

8) Which statements represents the number sentence $36 = 4 \times 9$?

A. 36 is 9 more than 4.

B. 36 is 9 times as many as 4.

C. 9 is 4 times as many as 36.

D. 9 more than 4 is 36.

9) A restaurant manager collected data on the lengths of time customers waited for their food. His data are shown in the dot plot. How many customers waited more than 20 minutes for their food?

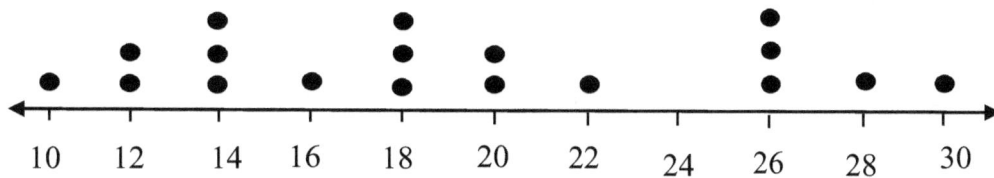

A. 5

B. 6

C. 12

D. 10

10) The Brandon family are on a 10-day road trip. They travel 11 hours each day for

4 days. They travel 6 hours each day for 9 days. How many hours does the

Brandon family travel during their road trip?

A. 98

B. 89

C. 94

D. 63

11) An unshaded fraction model is shown. Select the equivalent to the fraction model.

A. $\frac{3}{4}$

B. $\frac{6}{14}$

C. $\frac{3}{8}$

D. $\frac{5}{13}$

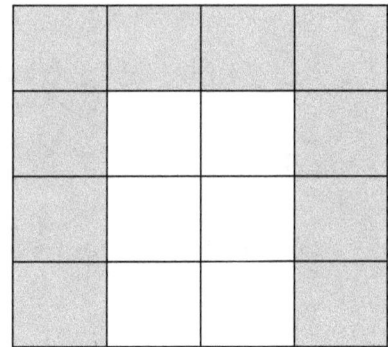

12) There are 20 books on a shelf. 8 of these books are new. What is the fraction of

all books is used books?

A. $\frac{2}{5}$

B. $\frac{8}{12}$

C. $\frac{3}{4}$

D. $\frac{3}{5}$

13) Which statement correctly compares the two values?

   A. The value of 6 in 3,169 is same as the value of the 6 in 4,635.

   B. The value of 6 in 3,169 is 10 the value of the 6 in 4,635.

   C. The value of 6 in 3,169 is 100 times the value of the 6 in 4,635.

   D. The value of 6 in 3,169 is 1000 the value of the 6 in 4,635.

14) The measure of angle ABC shown below is 135 degree. What is the measure of

   angle ABD?

   A. 60 degree

   B. 120 degree

   C. 30 degree

   D. 45 degree

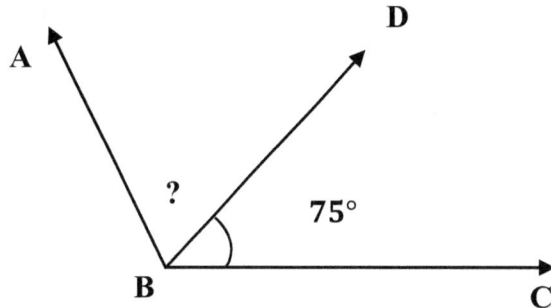

15) Kerry planted four types of beans and measured the height of the plants in

   centimeter after a week. Which type of bean has the shortest height?

   A. Velvet bean

   B. Lima Bean

   C. Bush Bean

   D. Soybean

Bean	Length (cm)
Soybean	$\frac{1}{5}$
Bush Bean	$\frac{1}{2}$
Lima Bean	$\frac{3}{4}$
Velvet bean	$\frac{9}{11}$

Session 2

❖ **Calculators are NOT permitted for this practice test.**

❖ **Time for the Session: 80 Minutes**

16) Emma has 105 pony beads. Miley has 15 times as many pony beads as Emma.

How many pony beads does Miley have?

A. 1,775

B. 1,757

C. 1,575

D. 1,577

17) Chris buys 20 candy bars at $1.50 each. If he sells them all in equal amounts for

the same price to five friends, how much money will each friend give Chris?

A. $5.2

B. $6

C. $7

D. $2.5

18) The table shows the relation between the number of hours Ian works each week.

Which could be a rule to find the hours (h) when given the week (w)?

Weeks (w)	3	4	5	6	7
Hours (h)	21	28	35	42	49

A. The output is $w - 21$

B. The output is $w \div 7$

C. The output is $w + 21$

D. The output is $w \times 7$

19) Andy has $400 to buy a new TV. $\frac{1}{5}$ of that money came from his grandmother

and he saved the rest. How much money did Andy save?

A. $80

B. $230

C. $320

D. $520

20) Mr. Hudson has several variable expenses that pays every month. The table

below shows the amount he paid for different expenses during last three months.

Which of following is his variable expenses?

Expense	April	May	June
Car payment	$290	$290	$290
Electricity bill	$82	$59	$63
Loan repayment	$710	$710	$710
Internet bill	$90	$90	$90
Parking	$32	$68	$43
Rent	$2,300	$2,300	$2,300

A. Parking and Electricity bill.

B. Rent and Parking.

C. Electricity bill, and Internet bill.

D. Car payment, Loan payment.

21) A school library has 72 new books. The books are divided into 6 shelves. Each shelf has n books. What equation can be used to solve for n books.

A. $n \div 6 = 72$

B. $72 - n = 6$

C. $n \times 6 = 72$

D. $72 + n = 6$

22) Adhira's family bring $10\frac{1}{3}$ liters of lemonade to a picnic. They drink $4\frac{5}{9}$ litters with lunch. Then they drink $3\frac{2}{9}$ litters with an afternoon snack. How much lemonade is left?

A. $2\frac{5}{9}$

B. $2\frac{4}{9}$

C. $1\frac{5}{9}$

D. $2\frac{2}{3}$

23) Aaron arrived at school at 6:50 A.M. He left school for 2 hours for a doctor's appointment. He was dismissed from school at 2:45 P.M. how long was at school?

A. 2 hours and 40 minutes

B. 4 hours and 50 minutes

C. 3 hours and 45 minutes

D. 4 hours and 35 minutes

24) What is the measure, in degrees of an angle that is equivalent to $\frac{1}{120}$ of a circle?

A. 3

B. 45

C. 120

D. 90

25) Which number line shows the correct locations of all given values?

18.30,19.80

A.

B.

C.

D.

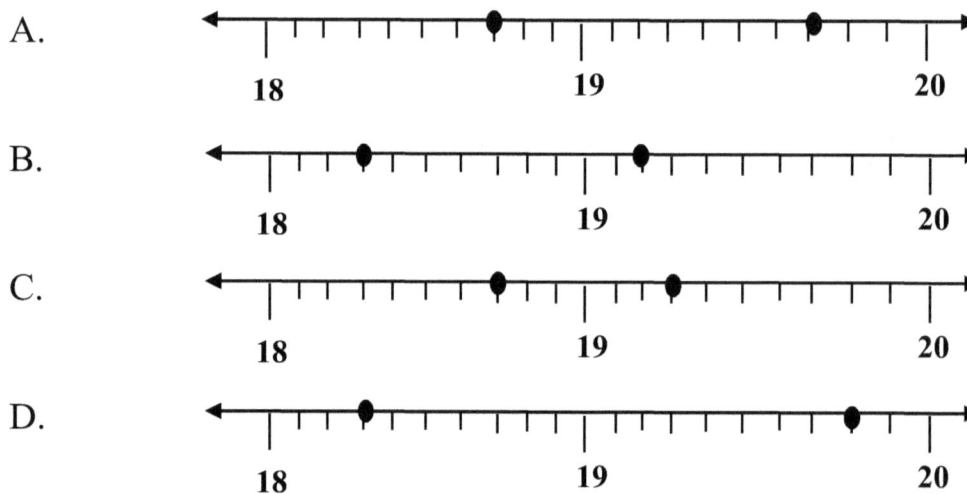

26) Bryson bought 4 shirts $18 each and 6 pants for $25 each. How much change did he get from $300?

A. $80

B. $78

C. $220

D. $150

27) What is the difference between 9.77 and 4.57?

A. $5\frac{1}{5}$

B. $6\frac{68}{100}$

C. $4\frac{19}{20}$

D. $5\frac{7}{8}$

28) Tina and her friend are buying lunch. They pay for 3 sandwiches and 3 drinks a $15 bill. The price for three drinks is $4. They received $5 in change. What was the price of each sandwich?

A. $4.2

B. $2

C. $2.5

D. $1.8

29) Riley bought 3.18 pounds of ice cream. She ate 0.33 pound. Her brother ate 0.45 pound. How much ice cream was left?

A. 0.78 pound

B. 2.80 pounds

C. 2.44 pounds

D. 2.4 pounds

30) There are 790 pounds fruits at the local fruit fair. Visitors buy a total of 341 pounds. Neighbors buy 214 pounds. The remaining fruits are donated equally to 8 charities. What are the greatest pounds of fruit that can be donated to each charity?

A. 39 pounds

B. 18 pounds

C. 29 pounds

D. 19 pounds

Answers and Explanations

Answer Key

Now, it's time to review your results to see where you went wrong and what areas you need to improve!

FSA Math Practice Tests

Practice Test 1

1	B	11	D	21	D
2	C	12	C	22	B
3	B	13	B	23	C
4	A	14	B	24	B
5	D	15	C	25	C
6	A	16	C	26	A
7	C	17	C	27	C
8	A	18	D	28	B
9	B	19	A	29	D
10	C	20	B	30	B

Practice Test 2

1	D	11	C	21	C
2	C	12	D	22	A
3	A	13	B	23	B
4	B	14	A	24	A
5	A	15	D	25	D
6	C	16	C	26	B
7	A	17	B	27	A
8	B	18	D	28	B
9	B	19	C	29	D
10	A	20	A	30	C

Practice Test 1

Answers and Explanations

1) Answer: B

We should have placed the 26 to the right of decimal point. The answer is 0.26.

2) Answer: C

perpendicular lines are lines that intersect (cross each other) at a right angle (90° angle). Then lines A and B appear to intersect at a right angle in the option C.

3) Answer: B

when 4-digit numbers are compared start with thousands place and then hundreds place, etc. If one whole number has a higher number in the thousands place, then it is larger than a whole number with fewer thousands. If the thousands are equal compare the hundreds, then the tens etc. when compared the numbers; 7 is between 6 and 8.

4) Answer: A

$$43 \times \frac{2}{6} = \frac{86}{6} = 14\frac{2}{6} = 14\frac{1}{3} \approx 15$$

5) Answer: D

$$
\begin{array}{r}
2,524 \\
\times \quad 5 \\
\hline
12,620
\end{array}
$$

6) Answer: A

68 and 62 is not divisible by 11 and 72 is bigger than 70, 66 has 8 factors. Then 66 is correct.

7) Answer: C

Use formula: 4 quarts = 1 gallons.

5 gallons 2 quarts + 3 gallons 3 quarts = 8 gallons 5 quarts = 9 gallons 1 quart

9 gallons 1 quart + 2 gallons 2 quarts = 11 gallons 3 quarts

8) Answer: A

I foot equals to 12 inches.

$8 \times 12 = 96$ inches, $96 \div 6 = 16$ barcelets

9) Answer: B

Sam: $3,600

Anna $= \frac{1}{20}$ Sam $= \frac{1}{20} \times 3,600 = \frac{3,600}{20} = 180$ dimes

George $= \frac{1}{5}$ Anna $= \frac{1}{5} \times 180 = \frac{180}{5} = 36$ dimes

Each dime is 10¢: $36 \times 10 = 360¢ = \$3.60$

10) Answer: C

Use fractions with same denominator: $\frac{6,300}{9} = \frac{6,000}{9} + \frac{300}{9}$

11) Answer: D

In the expanded form, we break up a number according to their place value and expand it to show the value of each digit.

$53,049.001 = (5 \times 10,000) + (3 \times 1,000) + (4 \times 10) + (9 \times 1) + (1 \times 0.001)$

12) Answer: C

Parallel lines are two lines that are always the same distance apart and never touch.

13) Answer: B

Area of field: $A_1 = 120 \times 55 = 6,600$

Area of tent: $A_2 = 18 \times 22 = 396$

All area without tent: $A = A_1 - A_2 = 6,600 - 396 = 6,204$

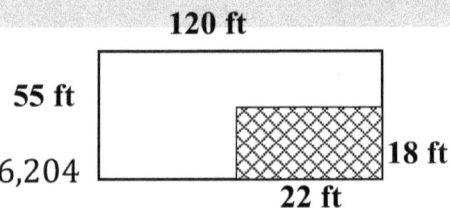

14) Answer: B

Every fraction has many equivalent fractions. $1 \times \frac{5}{6} = \frac{11}{6}$ and from the options provided:

A. $\frac{3}{5} + \frac{4}{5} = \frac{7}{5} \neq \frac{11}{6}$

B. $\frac{6}{6} + \frac{5}{6} = \frac{11}{6} = \frac{11}{6}$ is correct.

C. $\frac{1}{3} + \frac{1}{3} + \frac{2}{3} = \frac{4}{3} \neq \frac{11}{6}$

D. $\frac{1}{2} + \frac{1}{2} + \frac{1}{2} = \frac{3}{2} \neq \frac{11}{6}$

15) Answer: C

Russel: $6 \times \frac{1}{5} = \frac{6}{5} = 1\frac{1}{5}$; Philip: $13 \times \frac{1}{6} = \frac{13}{6} = 2\frac{1}{6}$

Compare $2\frac{1}{6}$ and $1\frac{1}{5}$: $2\frac{1}{6} > 1\frac{1}{5}$, then Philip has the biggest area of vegetables.

16) Answer: C

Look at the number in the tens' place and:

- If that digit is 0, 1, 2, 3, or 4, you will round down to the previous hundred.
- If that digit is 5, 6, 7, 8, or 9, you will round up to the next hundred.

$6,681 \cong 6,700$ and $6,721 \cong 6,700$

17) Answer: C

Quotient is the result when you divide one number by another.

dividend ÷ divisor = quotient; $71 \div 9 \cong 7$

18) Answer: D

When comparing decimals, start in the tenths place. The decimal with the biggest value there is greater. If they are the same, move to the hundredths place and compare these values: $24.45 < \boxed{} < 24.71 \Rightarrow 24.45 < 24.53 < 24.71$

19) Answer: A

A ray is a part of a line that has one endpoint and continues without end in one direction.

20) Answer: B

$112.38 + 96.53 + 86.24 = 295.15$

21) Answer: D

$11:23 + 1:45 = 13:08; 13.08 + 0:53 = 14:01$ P.M

22) Answer: B

To determine the measure of the angle, you should have found the two measures on the same scale (inside or outside) through which the rays of angle pass. Then, subtract the smaller measure from the larger measure.

On the inside scale: $130° - 70° = 60°$

On the outside scale: $110° - 50° = 60°$

23) Answer: C

You need to determine the total hours of rain for each day. 3 hours rain on Monday, 8 hours rain on Tuesday, $(8 - 4) = 4$ hours rain in Wednesday, $(3 \times 3) = 9$ hours rain in Friday, and $[30 - (3 + 8 + 4 + 9)] = (30 - 24) = 6$ hours rain in Thursday. Then you should have chosen the table with the same number of tally marks for each day.

24) Answer: B

$440 \div 11 = 40 price of each cake

Profit= Income – Cost: $40 - $25 = $15

25) Answer: C

Express a fraction with denominator 5 as an equivalent fraction with denominator 100 and use this technique to add two fractions with respective denominators 5 and 100.

$$\frac{3}{5} + - = \frac{80}{100} \rightarrow \frac{3}{5} + - = \frac{80}{100} \rightarrow \frac{60}{100} + - = \frac{80}{100} \rightarrow - = \frac{20}{100} = \frac{2}{10} = \frac{1}{5}$$

26) Answer: A

$60 \div 10 = 6$ or

You can count them.

1.	2.	3.	4.	5.	6.	7.	8.	9.	10.
11.	12.	13.	14.	15.	16.	17.	18.	19.	20.
21.	22.	23.	24.	25.	26.	27.	28.	29.	30.
31.	32.	33.	34.	35.	36.	37.	38.	39.	40.
41.	42.	43.	44.	45.	46.	47.	48.	49.	50.
51.	52.	53.	54.	55.	56.	57.	58.	59.	60.

27) Answer: C

$315 \div 7 = 45$

28) Answer: B

$33 + 52 + 44 = 129$ coins, then divide by 3: $129 \div 3 = 43$ coins.

29) Answer: D

$$4\frac{3}{4} - 2\frac{1}{4} = 2\frac{1}{2}$$

30) Answer: B

The line that divides a figure into identical halves is called the line of symmetry or the axis of symmetry. This shape has one line of symmetry.

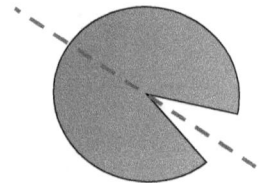

Practice Test 2
Answers and Explanations

1) Answer: D

The word form of thousand period is: 906, and digit in tens million and tens place are 0.

2) Answer: C

$$
\begin{array}{r}
410 \\
8\ )\overline{3{,}284} \\
\underline{32} \\
8 \\
\underline{8} \\
04 \\
\underline{0} \\
4
\end{array}
$$

3) Answer: A

To convert a Decimal to a Fraction, follow these steps:

- Step 1: Write down the decimal divided by 1, like this: $\frac{8.2}{1}$

- Step 2: Multiply both top and bottom by 10 for every number after the decimal point: $\frac{82}{10}$

- Step 3: Simplify (or reduce) the fraction: $\frac{41}{5}$ (divided by 2).

4) Answer: B

The perimeter of a rectangle is equal to the sum of all the sides.

Perimeter is: $30 + 30 + 18 + 14 = 92$ ft

Or $P = (30 + 30) + (18 + 18) = 60 + 36 = 96 - 4 = 92$

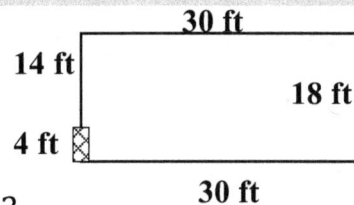

30 ft

14 ft

18 ft

4 ft

30 ft

5) Answer: A

Convert these fractions to equivalent fractions with a common denominator to compare them. Use the LCD to write equivalent fractions.

The LCD of (3, 5, 10, 12, 20, 30) is 60.

The equivalent fractions are: $\frac{2}{5} = \frac{24}{60}$, $\quad \frac{1}{12} = \frac{5}{60}$

$\frac{1}{30} = \frac{2}{60}$, $\quad \frac{1}{20} = \frac{3}{60}$, $\quad \frac{3}{10} = \frac{18}{60}$, $\quad$ and $\frac{2}{3} = \frac{40}{60}$

Then, $\frac{5}{60} < \frac{?}{60} < \frac{24}{60} \rightarrow \frac{5}{60} < \frac{18}{60} < \frac{24}{60}$

6) Answer: C

First count the amount of money he had before bought the book:

$4 \times \$7 = \$28$, $\qquad 5 \times \$0.05 = 0.25$, $\qquad$ and $12 \times \$0.01 = 0.12$

$\$28 + 0.25 + 0.12 = \$28.37$, then, subtract the price of the book:

$\$28.37 - \$20.68 = \$7.69$

7) Answer: A

Convert foot to inch: 8 feet $= 8 \times 12 = 96$, $96 + 12 = 108$ inches

Then, $108 \div 12 = 9$

8) Answer: B

36 is 9 times as many as 4.

9) Answer: B

The dot plot represented the same data below:

10, 12, 12, 14, 14, 14, 16, 18, 18, 18, 20, 20, 22, 26, 26, 26, 28, 30

There are 6 numbers greater than 20.

10) Answer: A

$4 \times 11 = 44$, and $6 \times 9 = 54$, then $44 + 54 = 98$ hours

11) Answer: C

The figure represents the fraction $\frac{6}{16}$, and the equivalent is: $\frac{3}{8}$ (Multiply by 2).

12) Answer: D

$20 - 8 = 12$ used book. Then, 12 books from 20 books means: $\frac{12}{20} = \frac{3}{5}$ (simplify)

13) Answer: B

The place value of 6 in 3,169 is tens.

The place value of 6 in 4,635 is hundreds.

To convert tens place to hundreds place we should have to multiply by 10.

14) Answer: A

$135° − 75° = 60°$

15) Answer: D

There are two main ways to compare fractions: using decimals or using the same denominator. In the decimal method, you should convert each fraction to decimals, and then compare the decimals.

$\frac{1}{5} = 0.2,$ $\frac{1}{2} = 0.5,$ $\frac{3}{4} = 0.75,$ and $\frac{9}{11} = 0.818$

Write the decimals in order from least to greatest: $0.2, 0.5, 0.75, 0.818$

16) Answer: C

$105 × 15 = 1,575$

17) Answer: B

$20 × 1.5 = \$30$, and $\$30 ÷ 5 = \$6$

18) Answer: D

Input - the act or process of putting in.

Output - production of a certain amount.

Input is week, and output is hours, then the hours is 7 times of weeks.

19) Answer: C

$\$400 × \frac{1}{5} = \frac{\$400}{5} = \$80$ then $\$400 − \$80 = \$320$

20) Answer: A

A variable expense is an expense or cost that occurs regularly, but the cost varies from month to month. Parking and Electricity bill as the only expenses that he paid different amount each month.

21) Answer: C

To find the number of books were in each shelf you have divided 72 by 6: $(72 ÷ 6) = n$, and by multiplication property: If you multiply both sides of an equation by the same number, the sides will stay equal to each other. (multiply by 6): $6 × n = 72$

22) Answer: A

$4\frac{5}{9} + 3\frac{2}{9} = 7\frac{7}{9}$, then, subtract: $10\frac{1}{3} − 7\frac{7}{9} = 10\frac{3}{9} − 7\frac{7}{9} = 9\frac{12}{9} − 7\frac{7}{9} = 2\frac{5}{9}$

23) Answer: B

First convert both times to 24 hours format, adding 12 to all pm hours, then subtract,

$2:45\ P.M = 14:45$; $14:45 - 6:50 = 13:105 - 6:50 = 7:55$

(If the start minutes are greater than the end minute: 1) Subtract 1 hour from the end time hours, 2) Add 60 minutes to the end minutes).

Then subtract 2 hours: $6:50 - 2:00 = 4:50$

24) Answer: A

There are 360 degrees in one complete circle around. $\frac{1}{120} \times 360° = \frac{360°}{120} = 3°$

25) Answer: D

First counted the number of sections on the number line between two numbers, since there are 10 section between 18 and 19 or 19 and 20, each section represented one-tenths. Then for 18.30 you count 3 sections after 18 and for 19.80 you count 8 sections after 19.

26) Answer: B

4 shirts $18 each: $4 \times 18 = \$72$

6 pants $25 each: $6 \times 25 = \$150$

$\$300 - (\$72 + \$150) = \$300 - \$222 = \$78$

27) Answer: A

$9.77 - 4.57 = 5.2$, Convert decimal to fraction: $5.2 = \frac{5.2}{1} = \frac{520}{100} = 5\frac{20}{100} = 5\frac{1}{5}$

28) Answer: B

$\$15 - (\$4 + \$5) = \$6$ paid for three sandwiches. Each sandwich: $\$6 \div 3 = \$2$

29) Answer: D

$0.33 + 0.45 = 0.78$ pound

$3.18 - 0.78 = 2.4$ pounds were left.

30) Answer: C

$341 + 214 = 555$, $790 - 555 = 235$

$235 \div 8 = 29.375 \cong 29$ pounds

"End"